Opportunities in Applied Environmental Research and Development

Committee on Opportunities in Applied Environmental Research and Development

Board on Environmental Studies and Toxicology

Commission on Geosciences, Environment, and Resources

National Research Council

NATIONAL ACADEMY PRESS
WASHINGTON, D.C. 1991

NATIONAL ACADEMY PRESS   2101 Constitution Avenue, N.W.   Washington, D.C. 20418

NOTICE: The project that is the subject of this report was approved by the Governing Board of the National Research Council, whose members are drawn from the councils of the National Academy of Sciences, the National Academy of Engineering, and the Institute of Medicine. The members of the committee responsible for the report were chosen for their special competencies and with regard for appropriate balance.

This report has been reviewed by a group other than the authors according to procedures approved by a Report Review Committee consisting of members of the National Academy of Sciences, the National Academy of Engineering, and the Institute of Medicine.

The National Academy of Sciences is a private, nonprofit, self-perpetuating society of distinguished scholars engaged in scientific and engineering research, dedicated to the furtherance of science and technology and to their use for the general welfare. Upon the authority of the charter granted to it by the Congress in 1863, the Academy has a mandate that requires it to advise the federal government on scientific and technical matters. Dr. Frank Press is president of the National Academy of Sciences.

The National Academy of Engineering was established in 1964, under the charter of the National Academy of Sciences, as a parallel organization of outstanding engineers. It is autonomous in its administration and in the selection of its members, sharing with the National Academy of Sciences the responsibility for advising the federal government. The National Academy of Engineering also sponsors engineering programs aimed at meeting national needs, encourages education and research, and recognizes the superior achievements of engineers. Dr. Robert M. White is president of the National Academy of Engineering.

The Institute of Medicine was established in 1970 by the National Academy of Sciences to secure the services of eminent members of appropriate professions in the examination of policy matters pertaining to the health of the public. The Institute acts under the responsibility given to the National Academy of Sciences by its congressional charter to be an adviser to the federal government and, upon its own initiative, to identify issues of medical care, research, and education. Dr. Kenneth I. Shine is president of the Institute of Medicine.

The National Research Council was organized by the National Academy of Sciences in 1916 to associate the broad community of science and technology with the Academy's purposes of furthering knowledge and advising the federal government. Functioning in accordance with general policies determined by the Academy, the Council has become the principal operating agency of both the National Academy of Sciences and the National Academy of Engineering in providing services to the government, the public, and the scientific and engineering communities. The Council is administered jointly by both Academies and the Institute of Medicine. Dr. Frank Press and Dr. Robert M. White are chairman and vice chairman, respectively, of the National Research Council.

Support for this project was provided by the Environmental Protection Agency, Grant No., 68D80079

Library of Congress Catalog Card No. 92-80241
International Standard Book Number 0-309-04698-X

Additional copies of this report are available from the National Academy Press, 2101 Constitution Avenue, N.W., Washington, D.C. 20418

S563

Printed in the United States of America
First Printing, October 1991
Second Printing, May 1992

Committee on Opportunities in Applied Environmental Research and Development

Richard N. L. Andrews, *Chairman*, University of North Carolina, Chapel Hill
David Bates, University of British Columbia, Vancouver
Richard A. Conway, Union Carbide Corporation, South Charleston
Donald Hornig, Harvard University, Cambridge
Duncan T. Patten, Arizona State University, Tempe
F. Sherwood Rowland, University of California, Irvine
Arthur C. Upton, New York University Medical Center, New York

Staff

Deborah D. Stine, Project Director
Robert B. Smythe, Program Director
Kathleen J. Daniel, Staff Officer
Ruth Crossgrove, Editor
Norman Grossblatt, Editor
Karen L. Hulebak, Senior Staff Officer
Bernidean Williams, Information Specialist
Boyce N. Agnew, Project Assistant
Ruth P. Danoff, Project Assistant

Sponsors

Environmental Protection Agency
Agency for Toxic Substances and Disease Registry, U.S. Public Health Service
National Institute of Environmental Health Sciences

Board on Environmental Studies and Toxicology

Gilbert S. Omenn, *Chairman*, University of Washington, Seattle
Frederick R. Anderson, Washington School of Law, American University, Washington, D.C.
John Bailar, III, McGill University School of Medicine, Montreal
Lawrence W. Barnthouse, Oak Ridge National Laboratory, Oak Ridge
Garry D. Brewer, Yale University, New Haven
Joanna Burger, Nelson Laboratory, Rutgers University, Piscataway
Yoram Cohen, University of California, Los Angeles
John L. Emmerson, Lilly Research Laboratories, Greenfield
Robert L. Harness, Monsanto Agricultural Company, St. Louis
Alfred G. Knudson, Fox Chase Cancer Center, Philadelphia
Gene E. Likens, The New York Botanical Garden, Millbrook
Paul J. Lioy, UMDNJ—Robert Wood Johnson Medical School, Piscataway
Jane Lubchenco, Oregon State University, Corvallis
Donald Mattison, University of Pittsburgh, Pittsburgh
Nathaniel Reed, Hobe Sound
F. Sherwood Rowland, University of California, Irvine
Milton Russell, University of Tennessee, Knoxville
Margaret M. Seminario, AFL/CIO, Washington, D.C.
I. Glenn Sipes, University of Arizona, Tucson
Walter J. Weber, Jr., University of Michigan, Ann Arbor

Staff

James J. Reisa, Director
Robert B. Smythe, Program Director for Exposure Assessment and Risk Reduction
David J. Policansky, Program Director for Natural Resources and Applied Ecology
Richard D. Thomas, Program Director for Human Toxicology and Risk Assessment
Lee R. Paulson, Manager, Toxicology Information Center

Commission on Geosciences, Environment, and Resources*

M. Gordon Wolman, *Chairman*, The Johns Hopkins University, Baltimore
Robert C. Beardsley, Woods Hole Oceanographic Institution, Woods Hole
B. Clark Burchfiel, Massachusetts Institute of Technology, Cambridge
Ralph J. Cicerone, University of California, Irvine
Peter S. Eagleson, Massachusetts Institute of Technology, Cambridge
Helen Ingram, Udall Center for Public Policy Studies, Tucson
Gene E. Likens, New York Botanical Gardens, Millbrook
Syukuro Manabe, Geophysics Fluid Dynamics Lab, NOAA, Princeton
Jack E. Oliver, Cornell University, Ithaca
Philip A. Palmer, E.I. du Pont de Nemours & Co., Newark
Frank L. Parker, Vanderbilt University, Nashville
Duncan T. Patten, Arizona State University, Tempe
Maxine L. Savitz, Allied Signal Aerospace, Torrance
Larry L. Smarr, University of Illinois at Urbana-Champaign, Champaign
Steven M. Stanley, Case Western Reserve University, Cleveland
Sir Crispin Tickell, Green College at the Radcliffe Observatory, Oxford, UK
Karl K. Turekian, Yale University, New Haven
Irvin L. White, New York State Energy Research and Development Authority, Albany
James H. Zumberge, University of Southern California, Los Angeles

Staff

Stephen Rattien, Executive Director
Stephen D. Parker, Associate Executive Director
Janice E. Mehler, Assistant Executive Director
Jeanette A. Spoon, Financial Officer
Carlita Perry, Administrative Assistant
Robin Lewis, Sr. Project Assistant

*This study was begun under the Commission on Physical Sciences, Mathematics, and Resources, whose members are listed in Appendix E, and completed under the successor Commission on Geosciences, Environment, and Resources.

*This report is dedicated to the late Norton Nelson,
former chairman of the Environmental Protection Agency's Science Advisory Board,
who proposed this study.*

Preface

There are many opportunities in applied environmental research and development. In investigating such opportunities, the committee attempted to look into the future; therefore, this report is not an exhaustive scientific report, but rather an attempt to summarize the opinions of experts in several fields. Four areas were investigated by this committee with the assistance of participants in a series of workshops: waste reduction, ecosystem and landscape change, anticipatory research, and long-term chemical toxicity.

One fundamental question is whether the Environmental Protection Agency and other agencies should fund only research linked specifically to its legislated regulatory objectives or whether it should also anticipate new or emerging, unregulated environmental problems. This committee believes that applied research should focus on the most serious environmental hazards rather than be driven by current regulatory priorities or news-media coverage.

Further, the committee believes that research is an essential foundation of environmental protection and that scientifically based advisory panels should play a larger and more visible role in fostering improvements in environmental science, such as developing high-quality research institutions and advising environmental agencies on ways to link research effectively with policy and management decision making.

The committee's efforts were facilitated by the many scientists and engineers who participated in its workshops. The committee wishes to thank them for their endeavors. In addition, the committee would like to thank Karen Hulebak and Kathleen Daniel, who began the project, and Deborah Stine, who guided the report through the review process, provided valuable comments, and diligently ensured that the document was complete. Others of the BEST staff who contributed to the efforts include James Reisa, director; Robert Smythe, program director; Lee Paulson, Ruth Crossgrove, and Norman Grossblatt, editors; Tania Williams, who produced the camera copy; and Boyce Agnew and Ruth Danoff, project secretaries.

Richard N. L. Andrews
Chairman

Contents

Page

INTRODUCTION 1

GENERAL CONCLUSIONS AND RECOMMENDATIONS 3

APPENDIXES
    A. Waste Reduction: Research Needs in Applied Social Sciences 11
    B. Measuring Change in Ecosystems: Research and Monitoring Strategies 97
    C. Research to Improve Predictions of Long-term Chemical Toxicity 127
    D. Research Needs in Anticipation of Future Environmental Problems 153
    E. Commission on Physical Sciences, Mathematics and Resources 173

Opportunities in Applied Environmental Research and Development

INTRODUCTION

This report provides an overview of several fundamental issues and priorities in applied environmental research and development. Specifically, it identifies detailed topics for investigation in the following four areas:

- Waste reduction.
- Ecosystem and landscape change.
- Anticipatory research.
- Long-term chemical toxicity.

Research topics in each area were developed in a series of four workshops during 1989; the selection of workshop topics and participants and the identification of themes and issues were carried out by the National Research Council's Committee on Opportunities in Applied Environmental Research and Development. This report, in part, attempts a look into the future—which not all persons see with equal definition and clarity. This report is a summary of the joint expert opinions of specialists in several fields. Only the first section of this document is the report of the committee; the appendices seek to capture the broader range of ideas suggested during workshop meetings.

As a matter of historical policy, research supported by the Environmental Protection Agency (EPA) has been linked closely to its mission. In 1981, the chairman and staff director of a congressional research oversight subcommittee expressed this principle as follows:

One of our fundamental premises is that EPA should conduct or fund only such research activities as will support its mission. That mission ... is well defined [by federal statutes], and the need is to translate legislated regulatory objectives into criteria for managing research.... The problems facing the agency exist now; so the question for EPA research managers becomes ... how to plan and operate a program that will be supportive of the immediate agency mission. (Brown and Byerly, 1981)

Although these authors acknowledged the need for more basic research and argued strongly against tying research too closely to the immediate needs of EPA's regulatory programs, the intent of Congress at that time was to restrict EPA's research program to support of its current regulatory agenda. That approach tended to prevent anticipation of new environmental problems and to rule out research on emergent environmental problems that were not yet the subject of regulatory attention, research on alternative policies and incentives for environmental protection other than those specified in existing regulatory statutes, and research on topics that were not advocated by EPA's regulatory program managers.

EPA's research program has also been heavily influenced by the exigencies of the annual budget process. Over the years, research has commonly been the EPA budgetary element that is funded last and reduced first, and research funding was cut severely over the past decade (EPA, 1988). This pattern of giving low priority to funding research has produced serious unpredictability and has disrupted the year-to-year continuity that is necessary to produce good research. Investigator-initiated requests for grants, a key instrument for supporting innovative research by the broader environmental-research community and for training new environmental professionals and researchers, must compete with the EPA's pressures to maintain its in-house laboratories. A substantial portion of the modest funding allocated to extramural research is often used to support contractors who work for EPA laboratories, rather than researchers in university-based programs. What increases have occurred in EPA's research budget have frequently been targeted for programs on high-visibility issues—such as acid rain, hazardous waste site cleanups, and the greenhouse effect. Although these issues are important, they are no substitute for a balanced and sustained research program.

In the spring of 1987, EPA Administrator Lee Thomas requested the EPA Science Advisory Board's (SAB's) advice on ways to improve strategic research planning in the agency. The board appointed a research strategies committee, which was chaired by former Deputy Administrator Alvin Alm and included five subcommittees on major research subjects. The committee completed its report in September 1988. This report, entitled *Future Risk: Research Strategies for the 1990s* (EPA, 1988), found that EPA's research priorities had been dominated by regulatory imperatives and pollution-control technologies. The committee concluded that future research priorities at EPA should be directed more broadly to the prevention or reduction of environmental risks (EPA, 1988): "EPA's R&D program has to be expanded and reoriented to include much more basic, long-term research not necessarily tied to the immediate regulatory needs of EPA's program offices."

Specifically, the SAB committee made 10 recommendations for EPA's research program, including new emphasis on pollution prevention and waste reduction at its sources, anticipation of new environmental problems (not just regulatory support for known problems), a national core program for ecological research and monitoring, improved understanding of human exposures to pollutants, epidemiological research, and improved education and training for environmental scientists and engineers and for society as a whole.

In September 1988, the National Research Council (NRC) established a Committee on Opportunities in Applied Environmental Research and Development at the request of EPA, the National Institute of Environmental Health Sciences (NIEHS), and the Agency for Toxic Substances and Disease Registry (ATSDR) of the U.S. Department of Health and Human Services. The committee's charge was to identify and evaluate important needs, opportunities, and capabilities in applied environmental research and development and to conduct up to four workshops on selected topics to augment the work of the research strategies committee of EPA's SAB.

The committee devoted substantial effort from the outset to identifying previous studies related to its charge and to selecting topics for workshops. Over the two decades of EPA's existence, numerous studies on environmental research needs have been carried out by the NRC, by other agencies, and by EPA itself (NRC, 1975a,b, 1977a,b,c,d, 1985, 1986; EPA, 1980, 1987, 1988; CEQ, 1985; GAO, 1979, 1987, 1988; NIEHS, 1988).

The present committee developed a list of over a dozen possible research topics for workshops. It also solicited suggestions from members of NRC commissions, boards, and staff, the sponsoring agencies (EPA, NIEHS, and ATSDR), EPA's Science Advisory Board and its research strategies committee, and other colleagues. In October 1988, the committee sponsored a public briefing by the SAB research strategies committee, to which it invited more than 50 other environmental scientists and representatives of business, government, academe, and other organizations to assist it further in identifying topics.

The committee eventually selected four subjects. Three were chosen initially for workshops: waste reduction (with emphasis on research needs in applied social sciences, engineering aspects having already been considered in recent workshops), ecosystem-level risk assessment (later called ecosystem and landscape change), and anticipatory research (to

---

[1] The BEST committee discussed environmental epidemiology at length, but did not choose it as a subject of a workshop. A previous NRC committee had recommended that EPA establish a program of epidemiological research as an element of long-term research on air pollution (NRC, 1985), and the research strategies committee reached a similar conclusion. The BEST committee considered carefully what additional recommendations it might make to strengthen EPA's presence in epidemiological research. It agreed with the finding of the earlier NRC committee and the Research Strategies Committee that a strong epidemiological research program is essential to EPA's mandate; this need was reiterated by

address strategies for using research to anticipate environmental problems and to set priorities for the future). A fourth topic, methods for testing for long-term chemical toxicity, was chosen after additional discussions with NIEHS and ATSDR.

The workshops were held over 9 months during 1989: ecosystem risk assessment, March 2-3, in Warrenton, Virginia; waste reduction, May 8-9, in Annapolis, Maryland; anticipatory research, June 14-15, in Woods Hole, Massachusetts; and chemical toxicity, December 13-15, in Washington, D.C. Each workshop involved the participation of the members of the committee and 20-40 additional persons with particular knowledge of the subject matter; participant lists are in Appendixes A-D. The draft report of each workshop was circulated to the workshop participants and members of the committee for review and comment (and ultimately reviewed through a peer review process).

When the workshops had all been held, the BEST committee developed overall conclusions about opportunities and needs in applied environmental research and development. Its conclusions are summarized in the next chapter, and summaries of the individual workshops appear in the appendixes that follow it.

GENERAL CONCLUSIONS AND RECOMMENDATIONS

On the basis of its workshops and its deliberations, the Committee on Opportunities in Applied Environmental Research and Development formulated five general conclusions and recommendations. More detailed suggestions for research priorities in particular fields of research appear in the appendixes that follow.

Research is an essential foundation for environmental protection: it is not a luxury to be deferred when budgets are tight, and it should not be conducted merely in support of existing regulatory programs.

Applied research can have at least four important functions in support of environmental protection and management. First, it can enable understanding and anticipation of environmental problems, identification of their causes, and estimation of the magnitude and importance of their effects. Second, it can help in generating new solutions to those problems through innovations in technologies and incentives for change in human behavior. Third, it can provide a basis for public policy; regulatory agencies must have evidence to justify proposed regulations to protect public health and the environment. Fourth, research can provide feedback on the effectiveness and impact of environmental policy and management, so that they can be corrected where necessary.

Applied environmental research has never been supported adequately by EPA. It has been disproportionately vulnerable to budget cuts and has suffered severe budget reductions (in real dollars) during the 1980s, even while EPA's mandates have expanded. Of the four functions that applied research can have, EPA has disproportionately emphasized its use in providing evidence to justify regulatory proposals. Environmental research must be more stably and adequately supported across all four of its functions.

Applied research should be directed toward the most serious environmental hazards, toward their root causes and the most promising opportunities to reduce them, and toward topics that are amenable to good applied research; it should not be driven merely by current regulatory priorities or news media coverage.

Regulatory priorities are inevitably shaped by laws, by judicial decisions and consent decrees, by

---

participants in the workshop on long-term chemical toxicity, discussed in Appendix C. The BEST committee decided for two reasons, however, not to sponsor a workshop on this topic. First, it believed that substantive research needs recommended by the earlier NRC committee were still valid and that questions of internal organization and structure aimed at implementing these recommendations could not be addressed usefully by a workshop of outside experts. Second, another NRC committee, the Committee or Environmental Epidemiology, has recently been established to provide a more systematic assessment of the whole field of environmental epidemiology and will undoubtedly develop more comprehensive recommendations than a single workshop could yield.

political salience and news media coverage, and by other factors that are not directly related to environmental or human health hazard. All too often, however, those same factors drive research priorities as well and are compounded by organizational imperatives, such as the maintenance of established research staff and program capabilities. One result is a tendency toward heavy investment in well-studied, continuing regulatory problems (such as criteria air pollutants), and in "crash" programs on new problems that attract political attention (such as acid rain and hazardous waste sites). The results often are serious underinvestments in research on other important questions that have not yet attracted as much attention (such as waste reduction, basic understanding of ecosystems, and long-term chemical toxicity).

Applied environmental research differs from basic research in that it attempts to understand and remedy practical problems. Its priorities should therefore pertain to the relative severity of such problems and the likelihood of solving them. However, this approach does not mean its priorities should be shaped primarily by current regulatory demands, because these demands themselves are rarely based either on systematic assessment of relative risks and opportunities or on any careful assessment of the likelihood that the results of research efforts would help to produce substantially better environmental conditions.

An effective program of applied environmental research would include a process for setting research and development priorities largely independent of immediate regulatory priorities. Priority setting would be guided by relative risks, amenability to research, and the quality of research proposed. It would also be guided by the need to provide a balance among the four functions of research identified above: to promote understanding, to find solutions, to inform policy, and to evaluate policy effectiveness. A process for setting research and development priorities that met those four goals well would provide the greatest possible overall benefit to regulatory programs, even though regulations directed at hazards that were less severe or less amenable to research solutions would receive less research attention than other topics.

Several research topics meet the criteria noted above and should be considered for addition to environmental agencies' research agendas. Topics identified by the committee include applied social science research (especially in the area of waste reduction) coordinated research of ecosystems and landscape change, anticipatory research program development, and improved methods for predicting long-term chemical toxicity.

Whatever the reasons—inadequate budgets, the momentum of existing research programs and personnel, pressures associated with regulatory priorities, resistance within EPA to new subjects of research, opposition from budget officials outside the agency, or some combination of these—important topics of applied environmental research have not received adequate attention. This failure constitutes a serious barrier to EPA's effective performance of its mission, and it must be remedied.

The subjects considered by the BEST committee illustrate the general problem.

• Waste Reduction. For 2 decades, EPA's primary tool for environmental protection has been the mandatory imposition of pollution-control technologies. Its research program therefore has been overwhelmingly devoted to the environmental transport and fate of pollutants, their health effects, and the development and assessment of pollution-control technologies. EPA has recently announced a fundamental change in emphasis, asserting that further progress in environmental protection will require systematic incentives for waste reduction (that is, for actions that will reduce or prevent the generation of waste materials and the use of energy through changes in input and production processes and through on-site recycling). The implementation of such incentives, however, requires not only changes in technological research priorities, but also research in the applied social sciences. It requires understanding of the economics of material and energy use, and of the human behavior patterns and influencing factors that drive economic patterns; and it requires empirical evaluation of the effects of policy incentives, both those intended to reduce waste and those that might inadvertently increase it. Understanding of environmental processes, health effects, and pollution-control technologies is also necessary, but by itself it is insufficient to meet the goal of environmental protection.

In other words, waste reduction must be rooted

more broadly in a systematic approach to human use of materials and energy, not just in improvements in engineering or regulatory efficiency. Most of the research needed to implement this new strategy remains virtually unaddressed by federal programs. One reason might be the relative novelty of waste reduction as a policy objective, but others include historical emphasis on regulatory support and pollution-control technologies, the absence of social-science research programs or staff expertise in the agencies, and a persistent prejudice against social-science research on the part of budget officials. If EPA is to be effective in encouraging market-oriented and other nonregulatory incentives for waste reduction, applied social-science research must be made an explicit and integral element of its agenda.

- Ecosystem and Landscape Change. Many of the most important environmental problems now facing the nation involve long-term, large-scale environmental degradation, such as the effects of complex chemical pollution, regional air pollution, coastal degradation, wetland and other habitat losses, and loss of biotic diversity. Ecosystem studies have existed for at least 2 decades, but so far the response of ecosystems and landscapes (large spatial units with interacting ecosystems) has been assessed in a piecemeal, fragmented manner that has usually failed to provide adequate estimates of uncertainties associated with environmental threats. The piecemeal character of the past studies leaves us with insufficient information to distinguish between human-caused perturbations that would have serious adverse consequences and perturbations that would be either beneficial or unimportant. Addressing these inadequacies requires a coordinated research program to identify key ecosystem-level characteristics of structure and function and to use them as environmental indexes to provide better information about the relationships between natural variability and anthropogenic perturbations in ecosystems. Such a program requires a core of common concepts, indicators, and standardized methods directed to the management objectives of ecosystem and landscape health and sustainability and an integrated, large-scale, long-term national program for regionally focused ecosystem monitoring, research, and risk assessment.

- Anticipation of Future Environmental Problems. While our ability to predict the future is limited, most current environmental problems have been known to scientists for some time. Nevertheless, EPA and society often react to environmental problems only when they become public crises. Timely research might have helped either to anticipate the crises or to provide more effective means to deal with them. However, many such problems are not amenable to direct, short-term attack, and require sustained anticipatory research to understand them and develop appropriate responses. Better baseline data, for example, would provide clearer understanding of the relationship between natural variability and human impacts. Earlier research on the effects of policy interventions would provide better understanding of the most effective responses. It is just that sort of research that has commonly been missing in research missions that emphasize support of current regulatory mandates, rather than exploration of emergent problems and trends. There is a serious need, therefore, for federal agencies to support research to identify and anticipate future environmental problems. Such research should include efforts to broaden the range of ideas and approaches in environmental research, and to increase interaction among the many disciplines relevant to environmental protection. It should also strengthen the connections between applied research and the incremental accumulation of more fundamental understanding. EPA has created such research units at least twice in its history—the Washington Environmental Research Center in the early 1970s and the Office of Exploratory Research in 1978—but in each case their success was limited. Budget cuts have had extremely detrimental effects throughout EPA's research program during the 1980s, but programs devoted to anticipating future problems seem to be especially vulnerable to budget cuts and to competition from research programs driven by more immediate pressures. The committee believes that such budget decisions represent false economy and that serious attention should be directed to the establishment of stable programs for research in anticipation of emergent environmental problems.

- Long-Term Chemical Toxicity. Humans are exposed to thousands of chemical compounds, a substantial but unknown number of which might have toxic effects in the concentrations at which people are exposed to them. Effects other than acute poisoning can be difficult to identify when they occur over long periods or after a period of latency. Dose-effect relationships (even of some chemicals that are known to be toxic at high doses) generally are not known well enough for confident assessment at the low doses characteristic of ordinary human exposure. Many compounds have not been well investigated for toxicity at all. Despite the major resources that have been devoted to chemical toxicity testing by government agencies and by corporations, there is a serious need to develop faster and more effective methods of predicting long-term chemical toxicity. Meeting this need will require better cooperation and coordination among investigators, disciplines, and data bases and across testing approaches; strategic planning to optimize the design of studies that use multiple testing methods; and development of more extensive experimental and epidemiological data on human exposures to validate models based on testing in other species. For example, a previous NRC committee recommended that EPA develop a long-term plan for research on air pollution and that population-based studies, in the form of a program in epidemiology, should be an integral part of that plan (NRC, 1985). The EPA Science Advisory Board's research strategies committee reached a similar conclusion (EPA, 1988), as did participants in the workshop on long-term chemical toxicity. Only through such a program can EPA attempt to validate its inevitably heavy reliance on extrapolation from laboratory-animal models by comparison with human data and thus reduce the uncertainty as to long-term impact of the chemicals on humans.

The most fundamental need in applied environmental research today, which transcends all specific topics, is to build a stronger science base—including research programs, traineeships, facilities, funding, and institutional arrangements, but especially people—for environmental protection.

The topics discussed above are important examples, not a comprehensive list, of the subjects on which both more and better research is needed to support environmental-protection decisions.

Others could easily be added—indoor air pollution, noncancer health hazards to sensitive populations, and methods for restoration of damaged ecosystems, for example—and new problems undoubtedly will continue to be identified. The great environmental priorities of the coming century—stabilizing human populations and developing sustainable relationships between their economic wants and needs and the ecosystems that support them—will require far more serious and creative research commitments than now exist, either in EPA or elsewhere.

The truly fundamental need, both pervading and transcending specific research topics, is for a commitment to environmental research itself—to improving our knowledge of environmental processes and human impacts on them. Society takes for granted the need for such research commitments in other important sectors, such as military technology, space exploration, commercially exploitable technology, and the prevention and cure of disease. One can no longer assume that environmental conditions favorable for human life and well being can be maintained without a comparable serious commitment to environmental research.

Four ingredients are essential for implementing an improved long-term environmental research and development program: high-quality research institutions, the production of highly qualified environmental scientists and engineers, adequate funding, and an advisory capacity to link research effectively with policy and management decision making.

- Research Institutions. The first need is for a set of institutional arrangements that promotes high-quality interdisciplinary and anticipatory research on environmental conditions, trends, and intervention strategies. Both EPA laboratories and universities have important barriers and disincentives to such arrangements, which must be addressed and corrected. Scientists in federal laboratories often are too narrowly engaged in generating data to support regulatory decisions and too isolated from the norms, ideas, and evaluations of the larger community of environmental scientists. University scientists often are too narrowly focused on questions of disciplinary interest and too isolated from the practical policy and management questions that their research might help answer. Both could benefit from increased support for interdisciplinary and cooperative research investigations, and from more vigorous use of existing mechanisms for interaction, such as Intergovernmental Personnel Act arrangements. In addition, more information should be exchanged among government, academia, industry, and other nongovernmental organizations.

- Environmental Scientists and Engineers. The second necessary ingredient is a community of highly qualified environmental scientists and engineers and a deliberate and effective strategy for maintaining it. Environmental research must be given sufficient priority, visibility, and support to attract the best of today's scientists and engineers—particularly the best young scientists and engineers—from all relevant fields of the natural and social sciences and engineering. There are good environmental scientists and engineers in the field today, but they are scarce in some key disciplines, and there is a serious general need for more first-rate talent devoted to environmental research. During the 1960s, for instance, before EPA was created, the U.S. Public Health Service provided strong and crucial support for university-based environmental research and professional training. It supported strong multidisciplinary research centers based at universities and, equally important, environmental-health traineeships to attract and support the next generation of promising environmental researchers and professionals. Many of today's leading environmental scientists, engineers, and other professionals are alumni of such traineeships. The only such program existing today, however, is limited to the research agenda of the National Institute for Environmental Health Sciences. NIEHS promotes high-quality interdisciplinary and anticipatory research. It is dedicated to high-quality, long-term research in its own laboratories, but, in contrast with EPA, it also provides stable research funding for extramural research. It also provides research and training support and visible and attractive career opportunities to talented young scientists, beginning with an excellent predoctoral and postdoctoral fellowship program. In the important field of health consequences of environmental chemical exposures, therefore, NIEHS exemplifies the sort of research commitment that is needed. But NIEHS's research agenda is directed primarily at biomedical research on the health effects of environmental toxicants. NIEHS' program in practice does not include all important environmental determinants of human health, let alone the broader areas of ecology, engineering, and human behavior that are important to EPA's mandate. A similar commitment is needed across the full agenda of environmental research and development needs.

- Research Funding. The third essential ingredient is adequate, dedicated, and continued funding—the necessary support for any program. As noted above, funding in the environmental field is far too little and is often vulnerable to interruption or redirection.

- Advisory Mechanism. The fourth critical ingredient is a means to assemble promising ideas continually—an advisory body through which emergent problems and strategic research priorities can be identified, important insights recognized and targeted for research support, and important findings disseminated to other scientists and to policy makers.

Scientifically based advisory panels should play a larger and more visible role in fostering improvements in environmental science.

A recent report on the mission and functioning of EPA's SAB, for example, noted that its overall purview is science for environmental protection, not merely ensuring the quality of the science base for regulation. Its current functions include reviewing regulatory science, research programs and the technical bases for various applied programs and advising EPA on infrastructural and technical management issues, emergencies, and broad strategic matters. The report recommended an increase of the SAB's functions to include providing scientific forums and pursuing outreach, advising on implementation and communication, and helping EPA to anticipate problems and act more strategically (EPA, 1989).

The Committee on Opportunities in Applied Environmental Research and Development urges that in addition to those specific functional changes, the SAB give explicit and continuing attention to the development of the overall science base of environmental protection. That includes not just science and other research within

EPA, but the broader field of the environmental sciences from which EPA's science programs must be drawn—setting research priorities, conducting quality evaluations, assessing the quality and needs of environmental-science education and training, encouraging institutional arrangements necessary for multidisciplinary environmental research, and identifying important barriers to the progress of applied environmental research and development.

The SAB is uniquely situated to provide oversight and leadership in this effort, and the BEST committee believes that such a mission would be an appropriate extension of its current mission and a major service to the field of environmental science.

REFERENCES

Brown, G.E. Jr. and R. Byerly, Jr. 1981. Research in EPA: A congressional point of view. Science 211:1385-1390.

CEQ (U.S. Council on Environmental Quality). 1985. Report on Long-Term Environmental Research and Development. Washington, D.C.: Council on Environmental Quality.

EPA (U.S. Environmental Protection Agency). 1980. Environmental Outlook 1980. Office of Research and Development. EPA-600/8-80-003. Washington, D.C.: U.S. Environmental Protection Agency.

EPA (U.S. Environmental Protection Agency). 1987. Unfinished Business: A Comparative Assessment of Environmental Problems. Office of Policy, Planning, and Evaluation. Washington, D.C.: U.S. Environmental Protection Agency.

EPA (U.S. Environmental Protection Agency). 1988. Future Risk: Research Strategies for the 1990s. Report of the Research Strategies Committee, Science Advisory Board. Report No. SAB-EC-88-040 and -040A-040E. Washington, D.C.: U.S. Environmental Protection Agency.

EPA (U.S. Environmental Protection Agency). 1989. The Mission and Functioning of the EPA Science Advisory Board. Final Report to the Board from the SAB Subcommittee on the Mission and Functioning of the Science Advisory Board. Washington, D.C.: U.S. Environmental Protection Agency.

GAO (U.S. General Accounting Office). 1979. Improving the Scientific and Technical Information Available to the Environmental Protection Agency in its Decision Making Process. CED-79-115. Washington, D.C.: U.S. General Accounting Office.

GAO (U.S. General Accounting Office). 1987. Air Pollution: EPA's Process for Planning, Budgeting, and Reviewing Research. GAO/RCED-88-47BR. Washington, D.C.: U.S. General Accounting Office.

GAO (U.S. General Accounting Office). 1988. Environmental Protection Agency: Protecting Human Health and the Environment Through Improved Management. GAO/RCED-88-101. Washington, D.C.: U.S. General Accounting Office. 246 pp.

NIEHS (National Institute of Environmental Health Sciences). 1988. Conference on Environmental Health in the 21st Century. Environ. Health Perspect. 86:175-309.

NRC (National Research Council). 1975a. Principles for Evaluating Chemicals in the Environment. Washington, D.C.: National Academy of Sciences. 454 pp.

NRC (National Research Council). 1975b. Decision Making for Regulating Chemicals in the Environment. Washington, D.C.: National Academy of Sciences. 231 pp.

NRC (National Research Council). 1977a. Analytical Studies for the U.S. Environmental Protection Agency: Perspectives on Technical Information for Environmental Protection, Vol. I. Washington, D.C.: National Academy of Sciences. 108 pp.

NRC (National Research Council). 1977b. Analytical Studies for the U.S. Environmental Protection Agency: Decision Making in the Environmental Protection Agency, Vol. II. Washington, D.C.: National Academy of Sciences. 249 pp.

NRC (National Research Council). 1977c. Analytical Studies for the U.S. Environmental Protection Agency: Research and Development in the Environmental Protection Agency, Vol. III. Washington, D.C.: National Academy of Sciences. 99 pp.

NRC (National Research Council). 1977d. Analytical Studies for the U.S. Environmental Protection Agency: Environmental Monitoring, Vol. IV. Washington, D.C.: National Academy of Sciences. 181 pp.

NRC (National Research Council). 1985.

Epidemiology and Air Pollution. Washington, D.C.: National Academy Press. 224 pp.

NRC (National Research Council). 1986. Ecological Knowledge and Environmental Problem-Solving. Washington, D.C.: National Academy Press. 388 pp.

Appendix A

Waste Reduction: Research Needs In Applied Social Sciences

A Workshop Report

Summary

How is it possible to reduce the amounts of waste that we generate and the environmental pollution that they cause, rather than continuing to spend vast sums to dispose of them, move them somewhere else, treat them, or clean up contaminated sites? A growing consensus now argues that it would be more effective and cheaper to prevent pollution and reduce wastes at their sources instead. The U.S. Environmental Protection Agency (EPA) recently undertook major new policy and research initiatives and created a new office to address these issues.

This report summarizes a workshop on applied social science research needs in waste reduction held May 8-9, 1989, in Annapolis, Maryland, under the auspices of the National Research Council's Committee on Opportunities in Applied Environmental Research and Development. The workshop was one of four held at the request of EPA, the National Institute for Environmental Health Sciences (NIEHS), and the Agency for Toxic Substances and Disease Registry (ATSDR). The committee organized the workshops with the goal of assessing the state of the science in each area and recommending long-term research needs and opportunities for advancing the state of the science. This workshop focused on specific research needs in measurement of waste reduction, institutional and behavioral barriers to waste reduction, and policy incentives for waste reduction, recognizing that considerable thought already has been given to technological research needs. It also addressed research needs for waste reduction in several specific nonindustrial sectors (agriculture, counties and municipalities, and municipal wastewater management). The workshop identified a serious deficiency in federal research attention to the societal (as opposed to technological) choices necessary for waste reduction and to the relative effectiveness of public policies that influence such choices.

WASTE REDUCTION

Waste reduction is defined by EPA to include industrial input substitution; product reformulation; process modification; improved housekeeping; and on-site, closed-loop recycling. It also includes "environmentally sound recycling" and waste-reduction opportunities beyond the industrial sector through individual behavior patterns, such as consumption or disposal habits, driving patterns, and on-the-job practices. In practice, however, EPA's waste reduction research program to date has emphasized engineering research aimed at the efficiency of industrial processes.

In the view of workshop participants, waste reduction must be rooted more broadly in a systematic approach to human use of materials and energy, not just in improvements in engineering or regulatory efficiency, nor in ad hoc campaigns to reduce particular materials or change particular behavior patterns. Participants noted that waste reduction is enhanced, for example, by such measures as reducing the number of processing steps between extraction and end use and the concomitant use of intermediate products such as packaging, increasing the life span and reusability of end products, conserving energy, and substituting more benign for highly toxic materials.

APPLIED SOCIAL SCIENCE RESEARCH NEEDS

In the view of workshop participants, the achievement of waste reduction requires not only research on low-waste technologies and their relative impacts on the environment, but also research on the human causes of waste and pollution and on the relative effectiveness of public policies intended to promote waste reduction and pollution prevention. Effective waste reduction cannot be achieved, for instance, without operational and consistent measurement concepts that are clearly linked to environmental protection purposes and to the realities of the production and consumption processes in which they are used. It cannot be achieved without systematic understanding of human behavior toward the environment, by individuals and by organizations, and of the factors that influence those behavior patterns either toward or against waste reduction and environmental protection. Finally, it cannot be achieved without empirical evaluation of the effects of public policy incentives, existing and proposed, toward advancing or retarding the achievement of environmental protection and other public policy goals. Understanding of environmental processes, health effects, and pollution control technologies is also necessary, but, without the information gained from applied social science research, this understanding will not be sufficient to meet waste reduction goals.

Three specific examples of such opportunities and needs in applied social science research identified by the workshop are (1) the measurement of waste reduction, (2) the investigation of institutional and behavioral barriers to it, and (3) the comparative study of public policy options for encouraging it.

Despite the importance of these social science issues, however, most of them remain virtually unaddressed by federal programs for applied environmental research and development. In the view of workshop participants, a key reason for this lies not only in the relative novelty of waste reduction as a policy objective, but also in the institutional barriers toward the types of research questions that exist within the field of applied environmental research, within the federal environmental research programs that support it, and within the research budget oversight process.

The research programs of EPA's Office of Research and Development and its laboratories, in particular, have focused almost exclusively on the technical aspects of environmental management—environmental fate and transport of pollutants, health and ecological effects, and control technologies. EPA's staff of research program managers and scientists reflects those disciplines and priorities. Some studies of policy effectiveness and economic impacts have been sponsored by EPA's Office of Policy, Planning, and Evaluation, but, in the view of workshop participants, these do not provide an adequate substitute for a coherent and adequately supported program on the social science and management aspects of waste reduction research.

To carry out an effective policy of waste reduction and pollution prevention, therefore, workshop participants concluded that it is important that federal environmental research programs make applied social science research on environmental management an explicit and integral element of their agendas. Participants asserted that sustained research on these social science issues will be just as important to the achievement of waste reduction and pollution prevention as will research on the more technical aspects of environmental science and technology. Important opportunities exist for the integrated pursuit of applied social science and technical research on environmental management, as well as a serious need to redress the relative lack of attention to the former.

MEASUREMENT

Participants agreed that, as a public policy goal, waste reduction must be defined and measured at the level of the individual waste generator or manager (business firm or operation, household or institution, urban jurisdiction, etc.) and across the aggregate of human involvement in processes of materials and energy extraction, conversion, use, and disposal. No single definition or measurement will serve all waste reduction purposes and needs; multiple measurements are needed. Workshop participants suggested the following specific research topics:

1. Identifying useful indicators and measurement units for each purpose and

identifying relevant differences in their applicability to industry, government, and other waste-generating sectors, such as minerals extraction, agriculture, commercial and institutional activities, and consumption;

2. Measuring the postproduction waste implications of major commodity chemicals and other products as incipient wastes over their life cycles and evaluating claims that particular products are "environmentally friendly";

3. Measuring the waste reduction effects of complex capital investment decisions, such as future wastes that are avoided, in addition to incremental reductions within existing production processes;

4. Relating plant-level measurements of waste reduction to the combined effects of waste reduction by multiple sources at regional and national scales and developing aggregate estimates of the quantities of waste that might prove reducible; and

5. Attempting to develop consensual definitions for terms related to waste reduction.

INSTITUTIONAL AND BEHAVIORAL BARRIERS

When presented with information showing the benefits of waste reduction, many waste generators show surprisingly little interest in change, even if their own apparent economic interests would be served. Important institutional and behavioral barriers to waste reduction appear to exist even within the business sector, and it seems likely that the perception of similar or additional barriers affects the behavior of other waste generators, such as agriculture, government and other nonprofit institutions, and households.

Institutional and behavioral research efforts on waste reduction are in their infancy, but workshop participants identified substantial bodies of related theoretical and empirical evidence that could be applied to this task. Such evidence includes business research on accounting and financial analysis methods; on management of technological innovations; on organizational behavior and corporate strategic decision making; and on the behavior of other organizations and individuals, including recycling, energy and water conservation, and others. Workshop participants suggested the following specific research topics:

1. Refining accounting rules to more accurately reflect waste management costs and value recovered materials and to define uniform minimum standards for disclosure of environmental impairment risks;

2. Developing methods for incorporating waste reduction into technological innovation and capital investment decisions and for identifying the advantages waste reduction offers in strategic corporate decision issues;

3. Systematically evaluating the anecdotal literature of waste reduction innovations to identify measures of success and major influences on waste reduction behavior (e.g., type of incentives, size of organization, type of technology, and organizational goals and internal structures); and

4. Identifying incentives that might be most effective in influencing waste reduction behavior outside the business production sector, such as by government agencies, nonprofit institutions (e.g., schools, hospitals, and universities), households, and others.

POLICY INCENTIVES

Governments are the source of many initiatives intended to encourage waste reduction, such as regulations and enforcement, taxes and subsidies, and information and technical assistance programs. Governments are the largest procurers of materials and energy in society as well as major generators of wastes and, in most communities, they are also the primary providers of waste collection and disposal services and can create additional incentives through the management and pricing of those services. Nonetheless, governments also provide incentives for other purposes that might conflict with waste reduction, such as subsidizing raw materials extraction.

A third important set of research needs, therefore, concerns the effects of public policy incentives on promoting or retarding waste reduction. Three particular needs were distinguished by the workshop participants: documentation of existing policy incentives regarding waste reduction, including legislative and administrative mandates, and comparative evaluation of their effects; refinement of the economics of waste management to include the implications of waste reduction; and implementation and compliance issues. Workshop participants suggested the following specific research topics:

1. Identifying existing government policies that encourage or discourage waste reduction behavior, including the effects of programs intended to promote waste reduction in the United States and abroad, and assessing the magnitude of their effects;

2. Identifying differences in policy incentives needed to affect the waste reduction behavior of small businesses, not-for-profit institutions, and households, as opposed to large business organizations;

3. Advancing understanding of the economics of waste reduction, especially by determining how economic analysis can be properly applied to waste reduction and used to compare it with alternative approaches to waste management;

4. Evaluating impacts of alternative waste reduction incentives on illegal dumping and other compliance problems; and

5. Identifying what types of educational programs or other incentives are effective in encouraging end users to make waste reduction a priority in their decisions regarding purchases and disposal.

NONINDUSTRIAL SECTORS: THREE EXAMPLES

Beyond the general research needs identified above, workshop discussion groups identified three types of waste sources outside the industrial sector to which additional research attention should be devoted: agriculture, counties and municipalities, and public wastewater treatment operations. Other important sectors were also noted, such as building construction and the packaging of consumer products, but given the limited time available, the groups devoted their efforts to these three sectors.

Agriculture

Often overlooked in the focus on industrial waste reduction, the agricultural sector is a large and important source of waste discharges and consequent environmental pollution, ranging from pesticides and fertilizers to soil erosion, salination, and animal wastes. The participants thought that, in principle, waste reduction in agriculture could be advanced by a shift from chemical- and energy-intensive agricultural practices, which combine high crop yields with high costs, to "low-input" agriculture, which combines somewhat lower crop yields with lower costs and more efficient use of chemical inputs. However, careful research is needed on the actual waste reduction benefits resulting from changes in agricultural practices. Workshop participants suggested the following specific research topics:

1. Investigating how well low-input agriculture is working—on what crops, at what scales, and with what effects on yields, costs, and environmental impacts compared with more intensive alternatives;

2. Identifying what factors have most strongly influenced farmers to adopt low-input practices and what policy incentives most effectively encourage these forms of agricultural waste reduction;

3. Identifying appropriate and effective waste reduction practices for animal and other operations and for new configurations of integrated farm businesses;

4. Identifying the waste implications of existing agricultural policies such as pesticide regulations, federal grading standards for agricultural products, grazing fees, and commodity price supports; and

5. Determining how farmers and other consumers (such as golf-course operators, parks departments, utility firms, and homeowners) actually use high-risk pesticides and fertilizers.

County and Municipal Governments

Local governments are not only regulators but also sources and primary managers of many waste streams, and as a group, they represent a large number of decision makers who face similar problems in managing wastes. However, they have not received the attention and research support for waste reduction that industry has. Most have relied heavily on traditional landfilling practices, the rapidly rising costs of which now make source reduction and recycling far more attractive as alternatives than in the past. Applied research is urgently needed, therefore, on policy and technological options for waste reduction by county and municipal governments. Workshop participants suggested the following specific research topics:

1. Developing generic protocols for identifying waste reduction opportunities for county and municipal governments, in their own operations and in the waste streams that they regulate and manage;
2. Evaluating how well existing local waste reduction initiatives are working: their effectiveness in reducing wastes, their cost, their applicability to larger scales or more general use, and other implications;
3. Evaluating public willingness to participate in new management programs, to comply with new costs and requirements, and more generally, to modify the material and energy intensities of the public's lifestyles; and
4. Evaluating the economics of waste reduction for local governments, including particularly the costs and benefits of alternative methods for waste reduction and the economics of marketing and procuring recovered materials by local governments.

Waste Reduction in Municipal Wastewater Management

Public wastewater treatment plants discharge wastes themselves and receive wastes from others. Through such instruments as pretreatment requirements, pricing policies, and sludge management activities, treatment plants have important opportunities to further waste reduction; but by the same token, they might also be affected by chemical substitutions and other waste reduction activities of their dischargers. Workshop participants suggested the following specific research topics:

1. Determining the effects of chemical substitutions and other waste reduction initiatives by dischargers on the operation of wastewater treatment plants and on the quantity and quality of wastewater sludges; and
2. Identifying new waste reduction opportunities in the management of municipal wastewater sludges.

The following report provides a more detailed accounting of the waste reduction research needs in applied social sciences identified by workshop participants.

Waste Reduction: Research Needs in Applied Social Sciences

INTRODUCTION

Background

This report summarizes a workshop on applied social science research needs in waste reduction held May 8-9, 1989, in Annapolis, Maryland. The workshop was one of four held under the auspices of the National Research Council's Board on Environmental Studies and Toxicology (BEST) at the request of the Environmental Protection Agency (EPA), the National Institute for Environmental Health Sciences (NIEHS), and the Agency for Toxic Substances and Disease Registry (ATSDR). BEST established the Committee on Opportunities in Applied Environmental Research and Development to organize the workshops with the goal of assessing the state of the science in each area and recommending long-term research needs and opportunities for advancing the state of the science. The other workshop topics were ecosystem and landscape change, research needs in anticipation of future problems, and research to improve predictions of long-term chemical toxicity. The workshop that developed this report identified a general need for applied social science research on waste reduction strategies, including specific research needs in measurement of waste reduction, institutional and behavioral barriers to waste reduction, and policy incentives for waste reduction.

A root cause of most environmental pollution problems is the emission of waste materials and energy (in economic terms, residuals) from human activities to the air, water, and land. Policies to control such pollution focused initially on "safe" disposal (increased dilution in air and water and sanitary landfills instead of open burning in dumps) and subsequently on waste treatment technologies, which changed the physical or chemical form of the waste materials before discharging them to the environment. The effect of such policies was sometimes to reduce the quantity or toxicity of some waste streams. More often, however, it simply displaced them to other places, future times, or other environmental media, used additional material and energy inputs in the treatment processes, and left other waste streams unchecked (Andrews, 1989). These problems were well characterized by researchers in the late 1960s but were not widely popularized until a decade or more later (Kneese and Bower, 1968, 1979; Royston, 1979).

At least as far back as 1968, researchers in environmental management proposed a hierarchy of waste management options in which waste reduction was identified as the preferred approach, before treatment or disposal (Kneese and Bower, 1968). EPA adopted this hierarchy of preferences in concept in 1976 (EPA, 1976); and the Hazardous and Solid Waste Act of 1984 contained an explicit policy directive "that wherever feasible, the generation of hazardous waste is to be reduced or eliminated as expeditiously as possible." In 1986, EPA authored a "waste minimization strategy" that advocated waste reduction but characterized it loosely to include all waste management approaches, including subsequent treatment, storage, and disposal as well as reduction at the source. This strategy also focused on materials legally defined as hazardous wastes and addressed only technical, rather than regulatory and economic, barriers to reduction (EPA, 1986).

In a review of the 1986 strategy document, EPA's Science Advisory Board (SAB) urged that

EPA take a broader view of waste minimization, not limited to hazardous wastes or even to substances traditionally viewed as wastes. The SAB recommended that this view include "any non-product substance that leaves a production process or a site of product handling or use." The SAB also urged special emphasis on "waste prevention (source reduction)," which it defined as a subset of waste minimization practices that includes in-process practices, as well as waste generation practices by product users and consumers that prevent or reduce waste generation per se (EPA, 1987a). The Congressional Office of Technology Assessment (OTA) also criticized the narrowness of EPA's approach (OTA, 1986, 1987).

In 1987-1988, the SAB sponsored a committee study on environmental research strategies for the 1990s, and the report of this committee included major emphasis on research needs for "risk reduction" (EPA, 1988a,b). This report urged that risk reduction be adopted as the central goal of EPA generally and of its research and development activities specifically; it reaffirmed waste prevention (source reduction) as the preferred strategy for risk reduction and urged that EPA develop a strong program of research related to questions in these areas that were unlikely to be undertaken by or to duplicate the research of the private sector. The report noted explicitly that such research should include technology-based strategies and strategies involving disciplines other than the physical and biological sciences and engineering. Examples of the latter included policy and economic incentives for risk reduction, risk communication and perception, environmental management and control systems, and education and training programs.

In January 1989, EPA issued a draft policy statement adopting pollution prevention through source reduction and "environmentally sound recycling" as an agencywide commitment and establishing a new Pollution Prevention Office to develop and implement this goal across all EPA programs (EPA, 1989a). Key components of this program are to include the creation of incentives and elimination of barriers to pollution prevention, efforts at cultural change emphasizing the opportunities and benefits of pollution prevention, and related research and educational activities. A pollution prevention research plan has also been prepared by EPA's Office of Research and Development. The deliberations of this workshop may thus assist EPA, as well as other federal and state agencies, in identifying important research topics as they pursue their pollution prevention initiatives.

Previous Studies by the National Research Council

Study committees of the National Research Council (NRC) have addressed related topics in several reports. A 1983 report on management of hazardous industrial wastes endorsed source reduction and noted the importance of nontechnical as well as technical factors but did not offer research recommendations on these topics (NRC, 1983). A 1985 report addressed institutional factors in reducing waste generation, but it was limited to hazardous wastes, and its research recommendations were limited to technological methodologies (NRC, 1985). Other recent studies on related topics include alternative agriculture (NRC, 1989), multimedia pollution control (NRC, 1987), and the use of mass-balance information in environmental management (NRC, 1990).

Workshop Focus

The purpose of this workshop was to provide recommendations on applied social science research needs and opportunities in the field of waste reduction. Recognizing the substantial numbers of similar workshops and reports already addressing engineering and industrial process research needs in this field, this workshop focused on research questions involving three other domains:

- The definition and measurement of waste reduction;
- Institutional and behavioral barriers to it; and
- Policy incentives for waste reduction.

All these domains are important, transcend particular industrial processes or user activities, are unlikely to be addressed adequately by the private sector alone, and might therefore be strong candidates for research attention by EPA and other agencies.

As a modus operandi, a set of six background papers was commissioned, and a substantial set of additional background materials was also distributed before the workshop. During the workshop, participants were divided into three working groups, each directed toward one of the three domains; the suggestions generated by each group were then discussed in plenary session. The charge to these groups was to identify as specifically as possible the questions related to their topic that deserve research attention, and why; to suggest methods by which these questions might usefully be approached; and insofar as possible, to recommend priorities among them.

In the course of their discussions, the workshop participants also generated suggestions in two other domains:

- Waste reduction in nonindustrial sectors (e.g., agriculture, municipalities, and municipal wastewater); and
- Research implementation issues, in particular the general need for more attention to applied social science aspects of waste reduction.

Additional suggestions were also provided by several participants after the workshop.

A draft report was prepared by the chairman, incorporating what appeared to be the most valuable research suggestions both from the workshop discussions themselves and from the background papers and postworkshop comments from participants. All participants were given the opportunity to review and comment on the draft report, and the revised report was further reviewed and approved by the sponsoring NRC committee.

A workshop is by nature primarily an idea-surfacing mechanism. It is not in the nature of such a process, especially one addressing a large and heterogeneous field, to generate an exhaustive or definitive list of all important research needs. This report does, however, reflect the best suggestions of a diverse and knowledgeable group of participants and is endorsed by the sponsoring committee; it will be a useful source of research ideas, examples, and priorities to the sponsoring agencies and to the larger research community.

WASTE REDUCTION

Current Definitions

The concept of waste reduction has now been widely adopted, and EPA has defined it specifically to include "industrial input substitution, product reformulation, process modification, improved housekeeping, and on-site, closed loop recycling" (EPA, 1989a).

EPA's proposed policy is specific to industrial waste reduction, although it does note that "individuals as well as industrial facilities or organizations can practice source reduction and recycling through changes in their consumption or disposal habits, their driving patterns, and their on-the-job practices."

At the same time, the EPA proposal acknowledged that "there are varying views among representatives of industry, public interest groups, state and local governments and others over the role of recycling in pollution prevention," and it invited comment on the appropriate role of environmentally sound recycling in its pollution prevention program. This variance among viewpoints is due in part to semantic confusion (waste reduction, source reduction, toxic use reduction, risk reduction, waste minimization, pollution prevention, etc.), in part to restrictive definitions (hazardous waste reduction versus all materials and energy, volume versus toxicity reduction), and in part to differences among the goals and perspectives of its interpreters.

From the perspective of many businesses, for instance, waste reduction means incremental steps taken in enlightened self-interest to increase the efficiency of use of materials and energy in their production processes (Royston, 1979). It might include, therefore, such activities as sale of waste materials as byproducts and on-site (not just closed-loop) reclamation of waste materials for recycling. In contrast, to some environmental groups, it means a far more fundamental effort to reduce the overall production and use of toxic chemicals, excess packaging materials, and other harmful or wasteful products (National Toxics Campaign, 1989). From EPA's perspective, it appears to mean a new synthesis of earlier policy reform efforts toward integrated pollution control across all environmental media, by means of economic incentives rather than mandated treatment technologies (EPA, 1989a).

Two examples illustrate the limits of current definitions of waste reduction with particular force. The first is the narrow emphasis to date on industrial process wastes, with little attention to other significant waste-generating sectors: minerals extraction, agriculture and forestry, water supply, energy conversion, military facilities, and product packaging and consumption, for instance. Waste reduction must be directed toward strategic priorities, not simply toward opportunities for anecdotal or localized success stories.

The second example is the inherently dissipative use of all energy and many materials: combustion, agricultural chemicals, detergents and cleaning agents, lubricants and solvents, paints and dyes, packaging materials, and others. These uses traditionally are not counted as "wastes" because they are intentional and economically valued, in contrast to the economically useless wastes defined more narrowly above. However, their environmental effects are qualitatively identical and often quantitatively greater. Waste reduction priorities have begun with obvious concerns such as hazardous industrial wastes and are now being extended to include discharges of unwanted materials to air and water, but sensible priorities must ultimately be based on the relative harm of all human discharges of materials and energy to the environment, whether as wastes or for economically valued purposes.

Basic Concepts

Waste reduction in principle, therefore, must be rooted not simply in incremental improvements in engineering or regulatory efficiency, nor in ad hoc campaigns to reduce particular materials, but in a systematic approach to understanding of human uses of materials and energy. These use patterns are simultaneously ecological and cultural acts, although the cultural incentives that drive them frequently fail to incorporate adequate acknowledgement of their ecological consequences.

All living things, including humans, are ultimately sustained by natural accumulations of materials and energy resources and by natural processes of dispersion and purification of the waste products. Every human use of materials or energy involves a process of extraction, materials processing, intermediate and final uses (sometimes recycling and reuses), and discard or dispersion back into the processes of the natural environment (see Fig. 1, p. 57). Most ubiquitous of these processes are respiration, by which humans and other organisms extract oxygen and emit carbon dioxide, and combustion, which produces large amounts of carbon and nitrogen oxides (and, in the case of fossil fuels in particular, reintroduces into the biosphere large quantities of carbon and nitrogen long stored out of circulation by natural processes). Also important are the extraction and use of water, petrochemicals, mineral ores, and plants and animals. Ayres, writing in another recent NRC report, recently characterized these processes as the "metabolism" of an industrial society (NRC, 1989).

Most materials pass through the economic system rather quickly, within a few months to a few years. In ecological terms, these processes can be defined as wasteful whenever they convert concentrated resources into dispersed, and thus less accessible, forms at a rate faster than these resources are reconcentrated by natural processes; the more rapid the conversion, the more wasteful are the cycles. They are especially wasteful when they cause cumulative degradation of those natural regeneration processes themselves (e.g., NRC, 1986, pp. 93-100). More than just "wasteful," such processes can cause serious adverse impacts on ecological processes and human health. Such practices often are characterized more colloquially as drawing down the capital rather than living off the interest of our ecological endowment or—to borrow another business metaphor—taking short-term profits rather than maintaining the capital "plant" that produces them.

From their producer's point of view, however, materials and energy become wastes in economic terms at any point in this cycle when the costs of repairing them (in the case of products), or of recovering, collecting, and transporting them for input to another use are greater than their value as inputs (Bower, 1989). They become wastes from society's point of view when their value as inputs to any other use is lower than the cost of discarding them to the environment. Key terms in this equation are the relative costs of alternative input materials (or energy sources); the real costs to society of environmental disposal (including not only management costs, but the opportunity costs of damaging other services the environment provides); and the fraction of those real costs that is actually charged to the waste producers. The underlying purpose of waste reduction must therefore be not just to pursue enlightened self-interest in production efficiency, but to reduce those human uses of materials and energy that have the most serious damaging effects on ecological regeneration processes. This purpose has two components: quantitative waste reduction, or reduction of the total discharges of materials and energy to the environment, and qualitative waste reduction, or beneficial substitution of less damaging materials for those wastes that have the most seriously disruptive impact on the environment when discharged. Both these components must be pursued, moreover, not merely through technological changes, but through changes in the whole range of cultural and economic influences that drive waste-generating or waste-reducing behavior. Waste reduction is enhanced, for example, by such initiatives as reducing the number of processing steps between extraction and end use and the concomitant use of intermediate products such as packaging; increasing the usable life span and reusability of end products; conserving energy; substituting more benign for highly toxic materials; and on the part of consumers, purchasing less-material-intensive and less-toxic products.

THE NEED FOR APPLIED SOCIAL SCIENCE RESEARCH

In the view of workshop participants, the achievement of waste reduction requires not only research on low-waste technologies and their relative impacts on the environment, but also research on the human causes of waste and pollution and on the relative effectiveness of public policies intended to promote waste reduction and pollution prevention. Effective waste reduction cannot be achieved, for instance, without operational and consistent measurement concepts that are clearly linked to environmental protection purposes and to the realities of the production and consumption processes in which they are used. It cannot be achieved without systematic understanding of human behavior toward the environment, by individuals and by organizations, and of the factors that influence those behavior patterns either toward or against waste reduction and environmental protection. Finally, it cannot be achieved without empirical evaluation of the effects of public policy incentives, existing and proposed, toward advancing or retarding the achievement of environmental protection and other public policy goals. Understanding of environmental processes, health effects, and pollution control technologies is also necessary, but, without the information gained from applied social science research, this understanding will not be sufficient to meet waste reduction goals.

Barriers to Waste Reduction Research

Despite the importance of these social science issues, however, most of them remain virtually unaddressed by federal programs for applied environmental research and development. In the view of workshop participants, a key reason for this lies not only in the relative novelty of waste reduction as a policy objective, but also in the institutional barriers toward the types of research questions that exist within the field of applied environmental research, within the federal environmental research programs that support it, and within the research budget oversight process.

EPA's draft research strategy for the 1990s, for instance, noted that agency's environmental research priorities to date have been tied closely to the mission of supporting the immediate needs of its regulatory programs and that each of those regulatory programs in turn has focused on only a single medium or problem in isolation (air pollution, water pollution, pesticides, etc.), on engineering control technologies or restriction of individual substances as solutions, and on regulations as policy incentives for the adoption of the control technologies.

Similar historical patterns, pursuing compliance with the same regulatory demands, have been evident in the programs of other federal agencies that sponsor or conduct applied environmental protection research, such as the departments of Energy and Defense.

"[These] treatment and control approaches have not been effective in achieving compliance," notes EPA's draft strategy, "and other approaches are needed" (EPA, 1989b, pp. 1-7). Developing such approaches will require important readjustments in applied research priorities as well as in management and regulatory programs.

Lack of Research Programs and Expertise

A serious barrier to the development of such new approaches, however, is the virtual absence within these environmental research programs of organizational units or staff expertise devoted to research on the social science aspects of waste reduction or of environmental management more generally.

EPA's Office of Research and Development (ORD) research programs and national laboratories, in particular, have focused almost exclusively on the technical aspects of environmental management—environmental fate and transport of pollutants, health and ecological effects, and control technologies—and its staff, research program managers and scientists, accordingly reflects those disciplines and priorities.

In recent years, some research studies on policy effectiveness and economic impacts have been sponsored by other units, chiefly within EPA's Office of Policy, Planning, and Evaluation (OPPE). These efforts have been largely limited to economic studies, however, and to supporting immediate regulatory needs as well. They do not yet address the broader range of research questions identified above, and are not in any case an adequate substitute for a coherent and adequately supported program devoted to the social science aspects of waste reduction research.

Commitment to Applied Social Science

To implement this research agenda, therefore, and to carry out an effective policy of waste reduction and pollution prevention, the workshop participants believe that it is essential that the federal applied environmental research programs, EPA's in particular, make applied social science research on environmental management an explicit and integral element of their research programs. To do this requires leadership and an organization (Hollod, 1989). It requires committing explicit resources, designating an organizational focal point, and staffing it with experts who can recognize and demand the same quality in these fields of research that EPA expects in its more familiar technical fields.

Whether this capacity should be in ORD, OPPE, or elsewhere; what its budget should be; and what balance it should strike between developing its own research staff and supporting extramural projects are issues to be resolved by EPA rather than by this workshop. Similar questions should also be addressed by the other agencies that have research interests in waste reduction and pollution prevention.

It is clear that sustained research on these sorts of questions will be just as important to the achievement of waste reduction and pollution prevention as will research on the more technical aspects of environmental science and technology. It is also clear that important opportunities exist for the integrated pursuit of social science and technical research on waste reduction, as well as a serious need to redress the relative lack of attention to the former.

The remainder of the report addresses three more specific areas of applied social science in which workshop participants identified particular waste needs—definition and measurement, institutional and behavioral barriers, and policy incentives—as well as three particular non-industrial sectors—agriculture, municipal waste management, and wastewater treatment—to which increased research attention might usefully be directed.

DEFINITION AND MEASUREMENT

Background

The lack of attention to the measurement of waste reduction poses an obstacle to attaining and measuring progress (Hirschhorn, 1989). As a public policy goal, waste reduction requires definition and measurement not only at the level of the individual waste generator or manager (business firm or operation, household or institution, urban jurisdiction, etc.) but also across the aggregate of human processes of materials and energy extraction, conversion, use, and discard. Without serious attention to this broader perspective, waste reduction initiatives might simply rearrange existing environmental management problems, without ensuring that the overall result will be better.

Measurement at the Micro Level

At the level of the individual organization, the measurements needed are those that will permit managers to incorporate waste reduction effectively to pursue organizational goals as well as to meet government reporting requirements. These measurements include, in general, what materials and energy are generated, identified by amounts and by the process and product from which they arise; what their environmental and associated legal, regulatory, and business risks are; what could be done to reduce them, including options for substitute materials, processes, products, and other ways of reducing them, each with its own associated costs and risks; and what the relevant manufacturing standards are, including efficiency measures and full costs (including disposal and liability) associated with each course of action (see, e.g., Hollod, 1989).

Even at this level, however, measuring waste reduction presents important challenges. One problem concerns how to define baselines and normalize measurements: for instance, waste generated (or reduced) on an absolute basis (in physical units), per unit of input materials (by weight or volume), or in economic units such as per dollar value of sales (see NRC, 1990). A second problem concerns how to account for displacement: reducing wastes by sending them to an off-site recycler, or even by incorporating them more efficiently into products on site, might or might not reduce the health or environmental impacts over the material's life cycle. Similar issues are raised by transformation and substitution effects: reducing one waste stream by substituting an alternative chemical or process will produce different waste streams, which may or may not be preferable to, or even commensurate with, the old ones.

In addition, much of the literature on waste reduction to date has addressed incremental changes in existing facilities. Many important decisions affecting waste generation, however, take place in decisions about capital investment in new plants and processes. Such decisions involve additional and, in some cases, different measurement challenges, such as accounting for nonexistent wastes that would otherwise have been produced (that is, wastes avoided rather than reduced), allocating joint costs and benefits involved in co-location of complementary facilities, and others. Given the influence of such investment decisions on waste streams far into the future, these measurement questions have considerable practical as well as theoretical importance.

Measurement at the Aggregate Level

From the broader perspective of public policy, additional measurement challenges arise. If we limit our attention to the perspectives of particular firms—or particular industrial, commercial, or institutional activities—or to wastes as a preconceived category, we are likely to set faulty priorities and to miscount as waste reduction actions that merely displace potential pollutants from one location or process to another (Andrews, 1989). Measurement research must also be directed to the aggregate waste effects of society's material- and energy-use patterns as a whole.

In particular, for example, a large and growing proportion of the total waste stream is attributable to products as wastes rather than manufacturing process wastes, especially given the considerable economic incentives that already exist to incorporate materials efficiently into products (Bower, 1989; Hirschhorn, 1989). The waste reduction issues and measurement questions for waste reduction related to products, however, might be quite different from those related to process waste reduction. Key considerations affecting products as wastes, for instance, include their usable life span, repairability, adaptability to other uses, recyclability, and environmental hazards.

In the absence of effective measurement and allocation of the costs of product waste disposal, market forces tend toward increasingly wasteful product characteristics: proliferation of more and more differentiated or specialized products, disposable rather than durable products, heterogeneous rather than pure materials, complex sealed and unrepairable component assemblies, smaller and smaller unit sizes of products (with associated increases in packaging wastes), and cosmetic specifications for product characteristics (for instance, paper brightness standards) that unnecessarily increase waste generation (Bower, 1989).

Wolf (1988, 1989) illustrates some of the measurement issues in evaluating process versus product wastes by comparing four firms that produce or use methylene chloride and shows the difficulties in attempting to impose a single scheme even of measurement, let alone of regulation, on all of them. Should we be more concerned about the 5% of methylene chloride that is lost during production or about the other 95% that is sold as product but then released to the atmosphere by users as a degreaser, a spray-paint propellant, or a foam-board blowing agent? If we seek to phase out all these uses of methylene chloride, will its replacements be safer? Will they be more hazardous perhaps in different ways, such as ozone depleters? Or will they be simply less known or less regulated "compliance chemicals?" The same questions can be asked of other waste reduction efforts and of course, most obviously, of toxins, such as pesticides, that are deliberately dispersed into the environment.

Measurement Issues

It is important, therefore, to identify what research is necessary to define and measure waste reduction at both these levels. This task requires consideration of such questions as the following (Andrews, 1989):

1. What are we trying to measure? Waste reduction is now an increasingly popular concept, but different users of it have different measurement needs. Are we trying to measure overall national progress in reducing waste or merely local progress in reducing discharges to local air, water, and landfills; to measure physical amounts of wastes reduced or reductions in toxicity and other adverse environmental effects; to measure the efficiency of a single industrial plant or to be able to compare across plants, products, or economic sectors? No single number is useful for all these purposes; multiple measurements are necessary.

2. What differences in measurements might be required in different types of decision units (e.g., extraction and agriculture, primary materials processing, secondary manufacturing and product formulation, packaging/container producers, and recycling/reuse businesses; offices, institutions, and public agency activities; large integrated firms versus small specialized firms)?

3. Is waste reduction best pursued and measured by targeting specific "high-risk" substances throughout their processes of extraction and use (e.g., chlorofluorocarbons, lead, and chlorine); by targeting particular stages of the waste generation process (extraction, manufacturing, commercial use, consumer use, and waste management); by targeting particular sectors, industries, or firms that are especially wasteful, especially hazardous, or especially attractive for opportunistic waste reduction; or by targeting product characteristics and specifications? What measurements would help to clarify these priorities?

Measurement Research Needs

Specific research suggestions identified by workshop participants include the following:

1. What are the most useful indicators and measurement units for measuring pollution control and waste reduction at the source? Answers are necessary in order to develop uniform strategies for waste reduction across the full spectrum of materials and energy transformation activities.

2. Can the same indicators and units be used across different industry types and sectors? Government activities, for instance, mirror many of the material and energy transformation processes of the private sector, yet are managed differently; industries differ in their opportunities and constraints for waste reduction. What differences must be taken into account, and how do these differences affect attempts to measure net or aggregate waste reduction? Empirical comparisons are needed to answer these questions.

3. Similarly, how should we measure waste reduction in sectors other than manufacturing, such as minerals extraction and processing,

agriculture and forestry, commercial and institutional use, and households?

4. How may one measure the postproduction waste implications of products as incipient wastes over their life cycles? Relevant considerations include increases or decreases in usable life span, recyclability, environmental impacts compared to their substitutes, and perhaps others.

5. How can one measure the waste reduction effects of complex capital investment decisions, such as future wastes that are avoided (as opposed to present wastes that are reduced), allocation of waste reduction benefits among multiple products and processes, and commensurability across qualitative changes in types of waste streams? This might be approached through empirical before-and-after studies, in plants being modified and in new plants, showing the overall shifts in waste streams that occur.

6. What are the full life-cycle waste implications of major commodity chemicals? Studies could be undertaken on waste reduction for five to ten priority chemicals throughout their uses and life cycles, including industrial and nonindustrial uses. Such studies would serve to refine measurement approaches and, more important, to identify the most important waste sources and pathways for high-priority environmental contaminants. Important issues include such questions as how to select the most important chemicals and how to collect quantitative data on chemical uses and losses (especially, for instance, dissipative uses).

7. What products are environmentally friendly? Manufacturers and environmental advocates are increasingly touting some products as less environmentally harmful than others; in Europe, this practice extends even to consumer guides and "blue angel" labels on recommended products. Some of these claims might be well founded; others might reflect simple or self-interested value judgments. Comparative case studies would be useful to illustrate the environmental impacts of selected products and their alternatives throughout their life cycles, from fabrication through ultimate disposal (for instance, paper versus foam cups, paper versus plastic grocery bags), and to demonstrate methodologically the full range of factors that should be examined to make such determinations.

8. What are the relationships between plant-level measurements of waste reduction and the combined effects of waste reduction by multiple sources at regional and national scales?

9. How can one estimate more accurately the quantities of waste that might prove reducible, as well as the time and risk implications associated with such reduction? Also, how do these estimates compare with the public's definitions and goals for acceptable progress in waste reduction?

10. Finally, can a voluntary dialogue among interested parties develop useful consensual definitions for the terms related to waste reduction? Such definitions are needed to specify what is to be measured in the first place.

INSTITUTIONAL AND BEHAVIORAL BARRIERS

Background

The slogan most widely used to promote waste reduction is that "pollution prevention pays," not just in societal terms but in the coin of direct self-interest to the waste generator. A growing list of anecdotes has been adduced to support this claim (e.g., EPA, 1987b; Huisingh et al., 1985; Sarokin et al., 1985; Royston, 1979; and Kneese and Bower, 1968).

Even when presented with information purporting to show benefits to self-interest, however, many waste generators evince surprisingly little interest in change. There appear to be important institutional and behavioral barriers to waste reduction that are not yet well understood even within the business sector, and it seems likely that the perception of similar or additional barriers affects the behavior of other waste generators, such as agriculture, nonprofit institutions, households, and others (Forbes, 1989).

Within a given organization, for instance, barriers to waste reduction might be perceived in various ways: as technological, financial, labor force related, regulatory, consumer related, supplier related, managerial, and perhaps others (Ashford et al., 1988). Concern about the cost and reliability of new technologies is always likely, and traditional accounting procedures often obscure the true costs of waste disposal

(Hirschhorn, 1989). In addition, implementing waste reduction can require changes in the physical processes and the human procedures of production, which can be organizationally disruptive or threatening and are therefore resisted: why disrupt established practices and product standards that are generally accepted by customers, suppliers, distributors, and trade associations?

At a far more general level, patterns of materials and energy use in every society are a matter of fundamental patterns of cultural perceptions and behavior. Such patterns are not immutable—energy conservation behavior changed dramatically in the 1970s, for instance, and recycling behavior changed in New Jersey as soon as it was required by law—but they are rooted in culture rather than merely in technology or economics, and must, therefore, be addressed more thoughtfully than simply through cost incentives.

In short, the ways in which people and organizations manage wastes are a matter not just of technology and costs but of "mind-sets": for example, lack of awareness or understanding of other options, lack of interest on the part of managers or senior executives, preference for familiar habits and patterns, absence of perceived rewards for "rocking the boat" with new ideas, apprehension about future regulatory expectations, perception of environmental protection as purely a compliance issue, and higher priority of other environmental concerns or mandates. At a more fundamental level, they are also matters of cultural and sometimes even religious values.

A second important area of research discussed in the workshop, therefore, is to identify how such barriers affect the behavior of the various types of waste generators and how they can most effectively be reduced. Put in more positive terms, what actions does the literature of institutional and organizational behavior suggest might facilitate changes to reduce waste generation, and what changes would be required to educate business executives, engineers, economists, laypeople, and others to adopt such changes?

Institutional and behavioral research on waste reduction is still in its infancy, but substantial bodies of theory and empirical evidence exist in related areas that could usefully be applied to this task. The literature includes research on accounting and financial analysis methods; the management of technological innovation; organizational goals and effectiveness; contingency theories of organizational behavior and corporate strategic decisionmaking; and a variety of more specific issues, such as the acceptance of innovations by production workers (and of agricultural extension recommendations by farmers), the role of internal incentives in improving business innovation (such as bonuses for cost-saving ideas and "quality circle" programs), and energy- and water-conservation behavior in response to the price and policy incentives of the 1970s.

Accounting and Financial Analysis

The workshop identified a major set of procedural barriers to waste reduction in the area of accounting practices, management information systems, and financial disclosure requirements.

First, under traditional accounting practices, waste generation measurements normally are not incorporated explicitly into the formal accounting/control framework: many companies have explicit rules forbidding the incorporation of such information. Some costs of waste disposal, such as sewer charges, are normally included in utility costs, whereas others might be treated as administrative costs or general overhead ("period costs"). In either case, many firms do not identify them in sufficient detail to charge them directly to individual products or processes, often because, despite their potential importance to environmental protection, they represent only a small fraction of total production costs. The result is that cost and risk assessments are separated from control and business-mix decisions, and managers are given no incentive to reduce waste disposal costs unless total margins (over all production lines) are threatened.

Second, even if a firm wishes to identify waste disposal costs in detail, practices vary widely, and no clear accounting standards are available concerning how to do this. One such problem, for instance, concerns how to allocate joint costs of waste disposal in multiproduct plants. Another is how to value materials and energy that are recovered, such as waste sawdust recovered from a sawmill and reused as an input to pulp manufacturing: Should it be priced at market value as though it had been purchased from another firm, as free because it was a waste, at negative cost because reusing it avoids disposal costs as well, or somewhere in between? Some analysts argue that recovery which is economically beneficial should not be counted as pollution control costs because the firm would presumably undertake it without additional public policy incentives or regulatory requirements, but some firms attribute the whole amount to pollution control costs. Significant differences might result in the apparent benefits of waste reduction, as well as in actual tax and cost considerations.

Third, important accounting issues arise in evaluating plant modernization or replacement decisions that serve in part to reduce waste discharges (to comply with water and air pollution standards, for instance) but also increase productivity and cut costs overall. Some firms have argued that the entire capital cost should be credited to waste reduction; some economists respond, however, that a large fraction of these costs would have been undertaken to increase overall economic efficiency in any case, and therefore, should not be credited to waste reduction.

Finally, traditional financial disclosure practices fail to recognize potential contingency costs of leaks, spills, unsafe waste disposal sites, and other environmental impairment liabilities until a lawsuit or regulatory action is initiated. As a result, managers fail to see the true financial risks of unsafe waste management (and the potential benefits of waste reduction), and capital markets might not distinguish appropriately between well-managed and risky firms (Todd, 1989; Naj, 1988).

Research is needed, therefore, to identify refinements in accounting practices that would incorporate waste reduction incentives more effectively into decisions by business executives or managers and by capital markets. A few initial attempts to do this are under development (see, for example, ICF Inc., 1988), but further and more systematic work is needed.

Management of Innovation

Within the business sector, waste reduction can also usefully be studied in the context of ongoing processes of technological innovation (Hollod, 1989). These processes are typically described as "life cycles," proceeding from initial breakthroughs in product innovation through later innovations in production processes (e.g., standardization, mass production), to a more rigid stage of stable, routine production, and ultimate displacement by other innovations—often from industries not wedded to the same core concepts or capital equipment (Baughn and Osborn, in press; Van de Ven et al., 1989; Burgelman and Maidique, 1988; Ashford et al., 1985; Caldart and Ryan, 1985; and Abernathy and Utterback, 1978).

In this context, fundamental reexamination of production processes is a rare event, one that normally occurs when new products are introduced or new plants are designed rather than during routine operations. Most changes in existing processes are more incremental—some large, some small—triggered by external factors such as increases in input prices, shifting demand, or regulatory requirements, and emphasizing marginal improvements in existing processes, procedures, and product characteristics.

If this view is correct, research on waste reduction opportunities might usefully start by distinguishing between organizational units that are most immediately involved in the design of new products and the construction of new plants or processes, and those only responsible for existing plants (Caldart and Ryan, 1985). The former might be more receptive to major waste reduction innovations and have important payoffs, inasmuch as the highly automated plants now being designed could be modified inexpensively before construction but, once constructed, might be far more costly to modify than the more labor-intensive processes they replace. The latter will likely be more receptive to incremental innovations.

Organizational Goals and Effectiveness

Even when pollution prevention apparently pays in direct economic terms, the literature of organizational behavior presumes that corporations have more than one goal and that their behavior in practice reflects political compromises among the multiple goals of their constituencies: owners, senior managers, employees, suppliers, distributors, customers, competitors, regulatory agencies, neighbors, the financial community, and society in general (cf. Scott, 1987; Cohen, 1984; and Cameron, 1980). Any attempt to intervene to achieve a particular outcome such as waste reduction, therefore, requires explicit consideration of how it advances or retards other corporate goals, what tradeoffs it requires among them, and how its pros and cons are perceived by the various affected constituencies. Research collaboration between behavioral scientists and other disciplines might be especially useful here, along with comparative studies across different firms, including those that have undertaken waste reduction initiatives.

As one example, in some business organizations, mid- to lower-level innovators have difficulty getting top-level support and approval for their proposals, while in other firms, innovative senior executives feel that most resistance to waste reduction innovation comes from mid- to lower-level managers. It would be useful to conduct careful comparative studies of the effectiveness of top-down versus bottom-up strategies for introducing innovative waste reduction practices.

Contingency Organization Theory

From a third perspective, waste reduction incentives can be viewed as a variety of contingencies intended to change organizations' behavior, and research can be directed to how economic organizations (or others) typically respond to such contingencies (Scott, 1987). This literature provides a rich source that might complement economic studies and policy analysis.

For instance, what elements of the organization's environment does a particular incentive change? Different mechanisms are transferred into the firm in different ways and can be expected to yield different types of internal reactions or performance changes: firms respond differently to regulatory requirements, to changes in production cost factors, to technical assistance from their trade associations, and to other mechanisms that might be used to encourage waste reduction.

Technical assistance programs are already a major element in both EPA and state programs for promoting waste reduction, for example, but these efforts would benefit from explicit research attention to how and through whom such assistance can most effectively be delivered. Agricultural extension programs provide one model; trade associations, another; university-based programs, government agencies, and consulting engineering firms, still others.

What kinds of organizations are the targets of waste reduction incentives? Large firms with rigid, well-known technologies might be expected to be more resistant to change than smaller firms with more flexible technologies; on the other hand, small firms often lack the expertise and the financial resources to experiment with new technologies for waste reduction.

What are the internal structures of the target organizations? Organizations with more levels of management, centralized decision-making systems, and departments organized by function (engineering, operations, marketing) are expected to respond more quickly to specific, large-scale threats, whereas organizations with fewer management levels, decentralized decisionmaking, and departments organized by division (product, territory, or consumer) are more likely to respond to subtle changes and incentives.

These questions and others suggest lines of inquiry that would provide better insight into effective means for promoting waste reduction.

Corporate Strategy

If waste reduction is to occur in business organizations, it must be linked to corporate strategic advantages and decision considerations, not merely to more mundane arguments that are of interest only to lower- and mid-level managers (see Miller, 1987; Porter, 1986). It may well prove that very different (and probably more long-term) incentives are important to achieving waste reduction at the level of corporate strategic planning than those that are most frequently considered at the level of particular existing products and processes.

These considerations include such questions as how heavily to weight waste and pollution considerations in future product and process investments, whether to locate new production facilities in the United States or overseas, how diverse a business mix the organization aspires to run (for instance, whether to integrate new production from waste to byproduct into the business), and whether and how to coordinate operations with other firms (for instance, co-location of complementary operations for cogeneration of energy, to shorten product sequences and reduce packaging or to exploit other waste reduction opportunities). For some firms, the enhanced reputation resulting from a recognized and effective waste reduction program might be a central driving force. For others, especially some small firms, technical and economic risk might overwhelm all other considerations, and demonstration of these factors in practice might be the most persuasive argument.

Behavioral Questions in Other Sectors

The research literature described above has dealt primarily with behavior in organizations, especially economic production organizations such as business firms. Additional questions arise, however, in attempting to promote waste reduction in other types of organizations, such as government agencies and nonprofit institutions, or by individuals and households. How, for instance, can one most effectively alter the behavior of pesticide consumers—commercial farmers, utilities, homeowners, local parks, golf-course operators, etc.—to reduce waste or substitute less-hazardous practices? What are the advantages and disadvantages of waste reduction to decisionmakers in large institutions and organizations outside the production sectors (e.g., hospitals, universities, multifamily housing, and shopping centers)? Will people reduce their use of materials in response to increased landfill user charges, or will they simply dump their trash illegally or burn it and thus increase air pollution? And finally, to what extent can organized public concern and media attention provide an effective incentive for waste reduction policymaking and implementation by governments?

Other bodies of literature might be useful for addressing these questions. Examples include studies of energy- and water-conservation behavior in the 1970s, the substantial literature that exists in the field of health behavior, the agriculture extension literature on acceptance of innovative practices by farmers, and the literature on compliance that links psychology and public policy research (see Boyer et al., 1987; Geller et al., 1982; Winkler and Winett, 1982).

Institutional and Behavioral Issues

To address these questions will require consideration of such issues as the following (Andrews, 1989):

1. Who reduces waste, who doesn't, and why? In the business sector, is interest in waste reduction more characteristic of particular types of firms (e.g., large versus small or resource extraction versus basic chemicals versus diversified consumer product manufacturing firms) or of firms with particular characteristics (accounting practices, leadership commitment, "corporate culture," research/innovation capability, etc.)? Outside the business sector, what differences are evident between those who actively reduce wastes and those who don't (institutions, government units, households, etc.), and why?

2. For each type of decision unit, what are the principal barriers to further reduction? Is it that technological options do not exist, are too costly, are not known, or might not be reliable? Or would such options conflict with other organizational objectives (short- and long-term profits, cost minimization, product characteristics and marketing strategies, convenience packaging, etc.)? What is the relationship among basic patterns of personal preferences, organizational norms, and cultural values? What research could assist in identifying these barriers more clearly and in evaluating interventions intended to reduce them?

3. What role do attitudes and perceptions (on the part of business managers, workers, senior executives, suppliers and distributors, consumers, etc.) play in promoting or retarding waste reduction? To what extent do these attitudes vary: for example, by type of business, by size and internal differentiation of firms, and by functional responsibilities within the firm (e.g., product design, manufacturing, sales, marketing, and environmental health and safety) and, in the consumer sector, by socioeconomic status or other factors? What incentives and communication strategies would most effectively promote waste

reduction behavior on the parts of these varied actors, and who would provide this information?

4. What other changes in business decision processes—for instance, in organizational design, accounting procedures, and internal incentives—might contribute to waste reduction, and what research might assist in evaluating these possibilities?

Institutional and Behavioral Research Needs

Specific research topics suggested by workshop participants include the following:

1. What changes could be made in full-cost product accounting rules to develop consistent definitions of waste management costs, to capture them at a level of detail useful for waste reduction decisions, to allocate them appropriately among products and processes, and to set consistent and appropriate economic values on recovered materials?

2. Similarly, what protocols could be developed to incorporate pollution costs and risk contingencies explicitly into the minimum disclosure standards of the Financial Accounting Standards Board (1975)?

3. How can waste reduction be incorporated most effectively into technological innovation and capital investment decisions? Research is needed (1) to identify those organizational units that are most immediately involved in conceiving, developing, and designing new products and production processes and (2) to develop incentives to incorporate waste reduction measures as fully as possible into their designs.

4. Is it more effective to create incentives to encourage waste reduction by promoting market penetration of new businesses and technologies or adaptation of stable or declining ones? For example, what would be the most effective strategies for achieving waste reduction in the production and use of agricultural chemicals? Sources of possible comparisons or analogies include the emergence of the biotechnology industry and the adaptations of cigarette manufacturing firms to declining U.S. markets. A particularly relevant and important example is the dramatic success in energy conservation achieved by some energy producers and users (industrial, commercial, institutional, and individual) in response to the major change in oil prices during the 1970s.

5. How does waste reduction advance or retard the achievement of corporate goals? More sophisticated and empirical work is needed on precisely how and for whom pollution prevention pays and under what circumstances, going beyond the simple balance-sheet calculations generally used so far to include other considerations important to the firm such as effects on existing operations and employees, suppliers, distributors, and consumer acceptance. Research collaboration between behavioral scientists and other disciplines might be especially useful here, as well as comparative studies across different firms in the same industries, including those that have and have not undertaken program initiatives for systematic waste reduction and comparing top-down and bottom-up strategies.

6. How does waste reduction behavior differ depending on the type of incentives created, the sizes of organizations and types of technologies that are their targets, the internal structures of those organizations, and other factors? There is now a growing anecdotal literature on waste reduction initiatives by many types of organizations of many different sizes; it would be timely to characterize and evaluate these experiences more systematically to discover what general principles and predictive hypotheses can be distilled from them.

7. What strategic advantages and issues does waste reduction pose to senior corporate executives? Research is needed to better characterize the relationships between waste reduction concepts and strategic corporate decision issues, as opposed to lower-level operational decision-making; the opportunities for waste reduction at that level of choices; and the advantages or disadvantages of those opportunities for the organization.

8. Finally, what differences in behavioral and institutional factors must be recognized in waste-generating sectors other than profit-making businesses, and what waste reduction incentives might be most effective in influencing behavior in these sectors? Examples include government agencies, nonprofit institutions (e.g., schools, hospitals, and universities), and households.

PUBLIC POLICY INCENTIVES

Background

Waste reduction is pervasively but inconsistently influenced by the behavior of governments. Governments are the source of many initiatives intended as incentives to encourage waste reduction: regulations and enforcement, taxes and subsidies, information and technical assistance programs, and others. In most communities, they are also the primary providers of waste collection and disposal services, and can create additional incentives through their management and pricing of those services. At the same time, governments also create incentives for many other purposes that could conflict with waste reduction, such as subsidizing raw materials extraction; the incentives they do create might have unintentionally adverse consequences for waste reduction; and governments are themselves the largest procurers of materials and energy in society, as well as major generators of wastes. These incentives include legislative and administrative mandates.

A third important set of research needs, therefore, concerns the effects of public policy incentives toward promoting or retarding waste reduction. Three particular needs can be distinguished: (1) documentation of existing policy incentives for or against waste reduction and comparative evaluation of their effects; (2) refinement of the economics of waste management to include the implications of waste reduction; and (3) implementation and compliance promotion.

Existing Policy Incentives

Many public policy incentives are already being used by individual states, localities, and nations to promote waste reduction (see, e.g., Bower, 1989; Curlee, 1989a; McHugh, 1989; Schecter, 1989). Regulatory approaches, for instance, include restrictions on the disposal of particular materials (e.g., hazardous wastes and plastic packaging); restrictions on the sale or use of particular materials (e.g., nonrecyclable containers and lead in gasoline); mandatory source separation and recycling programs for municipal solid wastes; and initiatives to standardize characteristics and labeling of products to facilitate recycling.

Economic incentives include raising landfill use charges and marginal cost pricing of waste disposal services; disposal taxes or "product charges" built into the costs of products such as packaging materials; deposit-refund or recycling credit ("ticket") systems; income and property tax credits, grants, or low-interest loans for investments in waste reduction and recycling facilities; and pricing preferences for recycled goods in government procurement policies. Many educational and voluntary programs can also be found; additional proposals—regulatory, economic, and informational—are rapidly being generated and proposed for legislation (OTA, 1989; Stavins, 1988).

Additional approaches are also under way in other nations, particularly in Europe and Japan, where densities of population and urban and industrial activities have long been higher than those now causing waste disposal problems in the United States (Curlee, 1989b). Examples include government control of all hazardous waste disposal ("flow control"); disallowing disposal of recyclable materials; prohibition of particular toxic materials, such as cadmium, in consumer products; promotional labeling of environmentally friendly products; and statutes holding manufacturers and distributors increasingly responsible for the disposal of their products as wastes (see Linnerooth and Kneese, 1989; Bothén and Fallenius, 1982).

In addition to these policy experiments intended to promote waste reduction, many existing public policies enacted for other purposes also influence waste disposal behavior, for and against waste reduction. Strict liability for hazardous waste cleanup, for example, is a powerful incentive for waste reduction, as are insurability requirements for environmental impairment liability; also, the increasing cost of landfill disposal of all wastes as old landfills fill up and new ones must be built to higher standards (and against public opposition) is rapidly increasing the relative attractiveness of waste reduction and recycling.

However, long-standing federal subsidies for mining, forestry, and agricultural operations continue to make extraction of virgin materials more attractive than recycled goods. Similar disincentives to waste reduction are built into many trade standards and procurement specifications—including many government quality-grading and purchasing specifications—dictating the material content or cosmetic qualities of products. Examples include paper brightness, meat marbling, "100% virgin materials," and others. The existing system of environmental protection regulations poses some obstacles to waste reduction (Hirschhorn, 1989), and many of the most important barriers are embedded in legislative policies rather than merely administrative regulations.

The first research need on policy incentives, therefore, is to identify systematically the policy incentives that have significant influence on waste reduction behavior in each economic sector and to document empirically their effects, magnitudes, and the sensitivity of waste generators' behavior to them (Curlee, 1989b). Whom do the policies affect, and how? How effective are changes in price incentives, standards, or regulations; labeling requirements; procurement preferences; and other policy incentives as stimulants to waste reduction? How sensitive are these effects to the magnitude of the incentive and to the time over which it occurs? How much might waste reduction be enhanced, if at all, by removal of existing policy disincentives to it? What combinations of incentives have demonstrated successes and failures, and why?

There is a growing anecdotal literature on some of these questions, which is much in need of more systematic empirical investigation, sensitivity analysis, and comparative evaluation. Belzer and Nichols (1988), for example, investigated economic incentives for hazardous waste minimization and used-oil recycling; they concluded that in the hazardous waste sector, substantial economic incentives already existed, but there was a natural lag time in their effects. However, in the case of used oil, the transaction costs involved in recycling might limit the effectiveness even of significant new economic incentives.

Economics of Waste Reduction

Second, many research questions deserve investigation to advance understanding of the economics of waste reduction.

From the perspective of an individual business or consumer, waste reduction is often worthwhile without any public policy incentives, simply out of personal moral ideals or self-interest. Many individuals and cultures value modest life styles and frugal use of materials and energy. Others find opportunities to increase efficiency by reducing the normal gap between the idealized model of rational economic behavior and the reality of unexamined assumptions, imperfect accounting practices, standard operating procedures, habits, imperfect knowledge of better practices, "lumpiness" of the capital costs of corrective measures, and other considerations.

Beyond these self-motivated actions, however, public policy incentives play a crucial role in the economics of waste management. Some pollution prevention initiatives that did not pay yesterday, for instance, do pay today, because the alternative—given today's environmental protection requirements—is either rising landfill charges or required control technologies. Some pollution prevention initiatives that still do not pay today might pay tomorrow because of tighter environmental protection standards, even more costly disposal charges, changes in liability doctrines, or other policy changes. Conversely, some pollution prevention that pays today might not pay tomorrow: for instance, if regulatory enforcement is relaxed (giving competitors a free ride), if new requirements require further or different reconfiguration of the same processes, if markets for recovered materials become glutted, or if today's waste reduction proves to have unanticipated adverse consequences of its own (Andrews, 1989; Curlee, 1989b).

Traditional environmental economics has developed primarily as a branch of welfare economics, concerned with the treatment of externalities as a basis for public regulation. Logical extensions of this work concern appropriate pricing of the impacts of waste disposal on nature's services and regenerative processes. Equally important questions, however, arise in the business economics of waste management, for waste generators and for recovered materials dealers; in municipal finance, such as the operation of recycling programs and impacts of waste reduction on capital investments in other waste management facilities; and in natural resource economics, such as the implications of waste reduction incentives for natural resource production, the extractive industries, and the values of natural resources in nonextractive uses.

For example, do current municipal waste disposal charges include the full costs of waste disposal and full benefits of waste reduction? How should such costs be calculated, and what differences would they make throughout the economy? Examples include basing disposal fees on the costs of past versus future facilities; imposing disposal charges or surtaxes on particular products; evaluating recycling programs by direct profitability or by avoided costs; and evaluating the job creation benefits of unskilled labor positions in waste separation and recycling jobs. What differences would changes in these practices make as incentives for waste reduction: to governments themselves; to businesses, other institutions, or households; to the basic materials processing industries; and to the achievement of overall waste reduction?

Similarly, could waste reduction be better effected by mechanisms to "charge back" disposal costs to waste generators or, farther back yet, to product designers and producers or extraction industries? What research would assist in evaluating the effectiveness and the potential side effects of such mechanisms? More generally, what would be the implications, economic and environmental, of different allocations of the costs of waste management (of which waste reduction is a part) among raw materials producers and manufacturers (and their investors), consumers, taxpayers, secondary materials producers, and future generations?

More broadly still, serious waste reduction might require deliberate government initiatives intended, for instance, to restrain the growth of fossil fuel use; to expand markets for recycled goods at the expense of extractive industries; to decrease the size of the packaging industry and of industries that trade heavily in toxic chemicals; to reduce the materials and energy intensity of economic production and consumption generally; and to encourage myriad other readjustments—many incremental and some, perhaps, fundamental—in overall patterns of materials and energy utilization (Ayres, 1989). Such initiatives would require careful analysis of both their environmental and their economic effects.

These examples only begin to suggest the range of economic research questions that must be addressed concerning waste reduction, by economists in collaboration with environmental scientists, public policy scholars, and others.

Implementation

Finally, research is needed on issues related to the implementation of public policy incentives for waste reduction. Chief among these are issues of compliance and enforcement. Little research has been done on environmental enforcement and compliance generally, and such studies are particularly necessary as governments embark on a variety of new policy initiatives intended to promote waste reduction.

These questions must be approached from economic and broader behavioral perspectives. Belzer and Nichols (1988), for instance, note that some popular forms of policy incentives, such as those that increase disposal costs, might not simply encourage waste reduction but also exacerbate illegal dumping and black markets in wastes; other types of incentives, such as deposit-refund systems, might create more effective and straightforward incentives for compliance. At the same time, waste reduction behavior involves more complex considerations than merely compliance with regulations and fees, and compliance itself involves more complex considerations than mere calculation of the "costs" of being caught, paying fines, spending time in jail, and so forth.

Research is needed, for instance, on the extent of noncompliance under various existing incentive regimes; on the characteristics of compliers versus noncompliers and explanations for the differences; on the role of enforcement as a policy incentive for waste reduction in itself and the factors influencing the effectiveness of existing enforcement programs; and on the operation of black markets in waste disposal and waste import or export.

Research is also required on the extent to which modest innovations in existing permitting and enforcement practices could yield significant improvements in the incentives for waste reduction. Joint permitting and joint enforcement procedures, for instance, cutting across the fragmented requirements of existing regulatory mandates, could provide clearer and more consistent signals as to what is required of a given facility, and thus promote more systematic management and reduction of waste streams than now occur.

Issues Concerning Policy Incentives

Research in this area might require consideration of such questions as the following (Andrews, 1989):

1. What policy incentives might be most effective for reducing particular high-priority categories of materials and energy use?

2. What existing public policies increase and decrease the incentives for waste reduction in each sector, and what research might assist in answering this question? Examples might include legislative mandates, regulations and regulatory uncertainty; enforcement practices and expectations; educational and technical-assistance services; nonenvironmental policies influencing business decisionmaking; liability standards and disclosure requirements; and taxes, subsidies, and stabilization measures differentially affecting raw and secondary materials markets.

3. What policy instruments appear to have the most effective positive influence on the waste reduction behavior of end users (individuals, households, retail businesses, institutions, etc.)? What instruments have been ineffective or create perverse incentives? Examples for investigation might include product-design requirements or disposal restrictions, availability of recycling services, deposit-return and ticket systems, waste end taxes, voluntary or mandatory source separation of recyclables, restrictions or charges on amounts of waste generated, and special charges or restrictions on more toxic products/wastes.

4. What research would best advance understanding of the economics of waste reduction?

5. What effects do compliance and enforcement practices have on waste reduction incentives?

Research Needs on Policy Incentives

Specific research topics suggested by workshop participants include the following:

1. What existing government policies tend to encourage and to discourage waste reduction behavior?

2. What have been the effects of existing programs intended to promote waste reduction in the United States and abroad? What policy options can one identify from these experiences, how effective are they, and what additional information must be collected to evaluate their effectiveness? To what extent are price or behavioral interventions alone, or combinations of the two, effective in encouraging waste reduction?

3. What differences in policy incentives are needed to affect the waste reduction behavior of small businesses, not-for-profit institutions, and households, as opposed to large business organizations, and which incentives are most effective?

4. What research would best advance understanding of the economics of waste reduction? What are the most economically promising opportunities for major waste reduction, and what kinds of waste management results would be expected from different levels of investment in waste reduction measures (e.g., marginal costs and time horizon)? How can the future costs and benefits of waste reduction initiatives be projected? More generally, how can economic analysis be properly applied to waste reduction and used to compare it against alternative approaches to management?

5. What are the impacts of alternative waste reduction incentives on compliance rates (e.g., illegal dumping, incineration by home owners, and black markets)? What policy incentives are especially effective in encouraging compliance?

6. What types of educational programs and other incentives are effective in encouraging end users to make waste reduction a priority in their decisions regarding purchases and disposal?

NONINDUSTRIAL SECTORS: THREE EXAMPLES

Beyond the general research needs cited above, the workshop identified three particular types of waste sources, all outside the industrial sector on which EPA's definition of waste reduction has focused, and to which additional research attention should be devoted: agriculture, municipalities, and public wastewater treatment operations. Other important sources were also recognized—for example, building construction and consumer product packaging—but given limited time, the workshop participants decided to devote their efforts primarily to these three sectors.

Research Needs in Agriculture

Though often overlooked in the focus on industrial waste reduction, the agricultural sector is a large and important source of waste discharges and consequent environmental pollution, ranging from pesticides and fertilizers to soil erosion, salination, and animal wastes. In principle, waste reduction in agriculture can be advanced by a shift from chemical- and energy-intensive agricultural practices, which combine high yields with high costs, to low-input agriculture, which combines somewhat lower yields with lower costs. In reality, there is a complex gradient of mixtures of practices between these two idealized stereotypes (NRC, 1989). Important research questions concern how well low-input agriculture is working in practice and how, if desired, it might best be encouraged.

1. How well is low-input agriculture now working: on what crops, at what scales, and with what effects on yields, costs, and environmental impacts compared with high-input alternatives? This research should lead to further conclusions concerning the generalizability of low-input practices to larger-scale usage and other crops and the adequacy and accuracy of existing information sources for farmers concerning the pros and cons of these practices.

2. What factors have most strongly influenced farmers to adopt low-input practices, and what policy incentives are most effective in encouraging these forms of agricultural waste reduction? In particular, to what extent do existing public policies (such as commodity price and farm income support programs, marketing orders, pesticide regulations, and food quality-grading norms) encourage or discourage agricultural waste reduction practices? Related needs are to develop better understanding of the transition process from high- to low-input practices and to identify mechanisms that would help farmers weather the costs and risks associated with this transition.

3. What waste reduction practices are appropriate and effective for animal or other operations and for new configurations of integrated farm businesses? The issues identified above have often been associated particularly with field and specialty crop production. Similar investigations should be directed to other operations, such as livestock and poultry production and fiber crops (cotton, trees), and to the potential for integrating diverse operations in new ways so as to reduce waste generation. Existing policy incentives that might influence such practices include federal meat-grading standards, grazing fees, and other agricultural policies.

4. Finally, how are farmers and other consumers (such as golf-course operators, parks departments, utility firms, and homeowners) actually using high-risk pesticides and fertilizers? It would be useful to have detailed studies of agricultural practices involving several of these substances that are arguably high-priority candidates for waste reduction efforts, including application and use levels, environmental settings in which they are used, control practices, and comparative costs and effects of alternative chemicals or practices.

County and Municipal Research Needs

Local governments are not only regulators but also sources and primary managers of many waste streams, and as a group, they represent a large number of decision-making units that face similar problems in managing wastes. Yet they too have generally lacked the degree of attention and research support for waste reduction that industry has received. Most have relied heavily on traditional landfilling practices, the rapidly rising costs of which now make source reduction and recycling far more attractive as alternatives than in the past (Melosi, 1981). Applied research is urgently needed, therefore, on policy and technological options for waste reduction in counties and municipal governments (Curlee, 1989a,b; EPA, 1989c; Prete et al., 1988; Sherry, 1988a,b). High-priority questions suggested by workshop participants include the following:

1. What generic protocols could be developed for identifying waste reduction opportunities for county and municipal governments, in their own operations and in the waste streams that they regulate and manage? This question might be

addressed through several case studies of specific localities; elements should include methods of waste stream characterization, of identifying waste reduction opportunities, of evaluating waste reduction incentives, and of incorporating waste reduction into municipal and county planning and management. Particular attention might be devoted in at least one such study to institutional waste streams (such as schools and hospitals), in addition to businesses and households.

2. How well are existing local waste reduction initiatives working? A growing body of anecdotes exists concerning policies and programs adopted by particular local governments, in the United States and elsewhere, to promote source reduction and recycling. Empirical research is needed to document and compare the effects of these initiatives: how effective are they in reducing wastes, what do they cost, what other implications do they involve, and how generalizable are they to larger-scale or more general use? Examples include disposal pricing and volume restrictions, deposit or ticket requirements, bans on the use or landfilling of particular materials, recycling incentives or requirements, and procurement and marketing programs for recycled materials.

3. Will people cooperate? A pivotal assumption in local waste reduction initiatives is public willingness to participate in new management programs, to comply with new costs and requirements, and more generally, to modify the material and energy intensities of their life styles. Examples of such initiatives include recycling programs and source separation requirements, household hazardous waste collection days, and volume-based disposal charges. Applied behavioral research is urgently needed to evaluate the validity and limitations of these assumptions, and to identify other changes in behavior patterns that may bring unanticipated benefits or problems. Some lessons may be available from past studies of energy- and water-conservation programs.

4. What are the costs and benefits of waste reduction to local governments? Economic research is needed on many questions related to waste reduction by local governments. Examples include the costs and benefits of alternative methods for waste reduction (and for waste management as a whole) to local governments and broader jurisdictions; the range of waste disposal costs under alternative schemes for waste reduction and management and associated efficiency and equity implications; and the economics of marketing and procurement of recovered materials by local governments. Comparative studies are also required concerning the relative effectiveness of waste reduction through private versus public service providers and at a municipal/county versus regional scale of service organization.

Municipal Wastewater Treatment Research Needs

Public wastewater treatment plants discharge wastes themselves and receive wastes from others. Through such instruments as pretreatment requirements, pricing policies, and sludge management activities, they have important opportunities to further waste reduction themselves, but by the same token, they may also be impacted either positively or negatively by chemical substitutions and other waste reduction activities of their dischargers (cf. Muir et al., 1989; Sherry, 1988c). A printer's switch from organic solvents to water-based inks, for instance, might on balance benefit waste reduction, but it might also cause a significant increase in chemical oxygen demand at the wastewater treatment plant.

1. What are the likely effects of chemical substitutions and other waste reduction initiatives by dischargers on the operation of wastewater treatment plants, and on the quantity and quality of wastewater sludges?

2. What new waste reduction opportunities exist in the management of municipal wastewater sludges? Could waste reduction be achieved by large-scale composting of solid waste? If so, would this be preferable to conventional land application of sludges alone?

The NRC Committee on Opportunities in Applied Environmental Research and Development endorses these ideas on applied social science research to address waste reduction issues. While recognizing that EPA is now developing a substantial research program on waste reduction and pollution prevention, the committee believes that these sorts of applied research, hitherto virtually unaddressed by federal environmental research programs, should be given serious consideration and support along with research on technological options. The committee hopes that the research needs proposed in this report will assist EPA and other federal and state agencies in identifying important research topics as they pursue their pollution prevention initiatives.

REFERENCES

Abernathy, W.J., and J.M. Utterback. 1978. Patterns of industrial innovation. Technol. Rev. 80:40-47.

Andrews, R.N.L. 1989. Research Needs for Waste Reduction. Background paper prepared for the National Research Council Workshop on Waste Reduction Research Needs, Annapolis, Maryland, May 8-9, 1989 (Board on Environmental Studies and Toxicology, Committee on Opportunities in Applied Environmental Research and Development). 16 pp.

Ashford, N.A., C. Ayers, and R.F. Stone. 1985. Using regulation to change the market for innovation. Harv. Environ. L. Rev. 9:419-466.

Ashford, N., A. Cozakos, R.F. Stone, and K. Wessel. 1988. Design of Programs to Encourage Hazardous Waste Reduction: An Incentives Analysis. Project No. P21013. Trenton: Division of Science and Research, New Jersey Department of Environmental Protection.

Ayres, R. 1989. Industrial metabolism. Pp. 23-49 in Technology and Environment, National Academy of Engineering, J.H. Ausubel and H.E. Sladovich, eds. Washington, D.C.: National Academy Press.

Baughn, C., and R. Osborn. In press. Strategies for successful technological development. J. Technol. Transfer.

Belzer, R.B., and A.L. Nichols. 1988. Economic Incentives To Encourage Hazardous Waste Minimization and Safe Disposal. Prepared for the U.S. Environmental Protection Agency under Cooperative Agreement No. CR813491-01-2. Washington, D.C.: Office of Policy, Planning, and Evaluation, U.S. Environmental Protection Agency.

Bothén, M., and U.-B. Fallenius. 1982. Cadmium: Occurrence, Uses, Stipulations. SNV PM 1615. Solna: National Swedish Environmental Protection Board.

Bower, B.T. 1989. Economic, Engineering, and Policy Options for Waste Reduction. Paper prepared for the National Research Council Workshop on Waste Reduction Research Needs, Annapolis, Maryland, May 8-9, 1989 (Board on Environmental Studies and Toxicology, Committee on Opportunities in Applied Environmental Research and Development). 24 pp.

Boyer, B., E. Meidinger, J. Thomas, and J. Singh. 1987. Theoretical Perspectives on Environmental Compliance. Prepared for the U.S. Environmental Protection Agency under Contract No. 68-01-7252. Washington, D.C.: Regulatory Innovations Staff, Office of Policy, Planning, and Evaluation, U.S. Environmental Protection Agency.

Burgelman, R.A., and M.A. Maidique. 1988. Strategic Management of Technology and Innovation. Homewood, Ill.: Irwin. 604 pp.

Caldart, C.C., and C.W. Ryan. 1985. Waste generation reduction: A first step toward developing a regulatory policy to encourage hazardous substance management through production process change. Hazard. Waste Hazard. Mat. 2:309-331.

Cameron, K. 1980. Critical questions in assessing organizational effectiveness. Organ. Dyn. 9:66-80.

Cohen, M.D. 1984. Conflict and complexity: Goal diversity and organizational search effectiveness. Am. Polit. Sci. Rev. 78:435-451.

Curlee, T.R. 1989a. Source Reduction and Recycling as Municipal Solid Waste Management Options: An Overview of Government Actions. Prepared for U.S. Department of Energy under Contract No. DE-AC05-84OR21400. Oak Ridge, Tenn.: Oak Ridge National Laboratory. 31 pp.

Curlee, T.R. 1989b. Source Reduction: Potential Research Issues. Paper prepared for the National Research Council Workshop on Waste Reduction Research Needs, Annapolis, Maryland, May 8-9, 1989 (Board on Environmental Studies and Toxicology, Committee on Opportunities in Applied Environmental Research and Development). 2 pp.

EPA (U.S. Environmental Protection Agency).

1976. Effective Hazardous Waste Management (Non-Radioactive); Position statement. August 18. Fed. Regis. 41(161):35,350.

EPA (U.S. Environmental Protection Agency). 1986. Report to Congress: Minimization of Hazardous Waste. EPA/530-SW-86 033A. Washington, D.C.: Office of Solid Waste, U.S. Environmental Protection Agency. Available from NTIS as PB87-114336.

EPA (U.S. Environmental Protection Agency). 1987a. Review of the Office of Research and Development's Waste Minimization Strategy. Report of the Environmental Engineering Committee of the Science Advisory Board. SAB-EEC-88-004. Washington, D.C.: U.S. Environmental Protection Agency.

EPA (U.S. Environmental Protection Agency). 1987b. Waste Minimization: Environmental Quality With Economic Benefits. EPA/530-SW-87-026. Washington, D.C.: Office of Solid Waste and Emergency Response, U.S. Environmental Protection Agency.

EPA (U.S. Environmental Protection Agency). 1988a. Future Risk: Research Strategies for the 1990s. SAB-EC-88-040. Washington, D.C.: Science Advisory Board, U.S. Environmental Protection Agency. 19 pp.

EPA (U.S. Environmental Protection Agency). 1988b. Appendix E: Strategies for Risk Reduction Research. Report No. SAB-EC-88-040E. Washington, D.C.: Science Advisory Board, U.S. Environmental Protection Agency.

EPA (U.S. Environmental Protection Agency). 1989a. Pollution Prevention Policy Statement. January 26. Fed. Regis. 54(16):3845-3847.

EPA (U.S. Environmental Protection Agency). 1989b. Protecting the Environment: A Research Strategy for the 1990s. April 1989 draft. Washington, D.C.: Office of Research and Development, U.S. Environmental Protection Agency.

EPA (U.S. Environmental Protection Agency). 1989c. The Solid Waste Dilemma: An Agenda for Action. Final report of the Municipal Solid Waste Task Force. EPA/530-SW-89-019. Washington, D.C.: Office of Solid Waste, U.S. Environmental Protection Agency. 70 pp.

FASB (Financial Accounting Standards Board). 1975. Accounting for contingencies. Financial Accounting Standards No. 5. Pp. 732-751 in Financial Accounting Standards Board Statements. Stamford, Conn.: Financial Accounting Standards Board.

Forbes. 1989. The best way to waste not. Forbes 154(12):20.

Geller, E.S., R.A. Winett, and P.B. Everett. 1982. Preserving the Environment: New Strategies for Behavior Change. New York: Pergamon. 338 pp.

Hirschhorn, J. 1989. Research Issues and Needs for Waste Reduction. Paper prepared for the National Research Council Workshop on Waste Reduction Research Needs, Annapolis, Maryland, May 8-9, 1989 (Board on Environmental Studies and Toxicology, Committee on Opportunities in Applied Environmental Research and Development). 2 pp.

Hollod, G.J. 1989. Perspective on a National Waste Reduction Initiative. Paper prepared for the National Research Council Workshop on Waste Reduction Research Needs, Annapolis, Maryland, May 8-9, 1989 (Board on Environmental Studies and Toxicology, Committee on Opportunities in Applied Environmental Research and Development). 6 pp.

Huisingh, D., H. Hilger, S. Thesen, and L. Martin. 1985. Profits of Pollution Prevention: A Compendium of North Carolina Case Studies in Resource Conservation and Waste Reduction. Raleigh: Pollution Prevention Pays Program, North Carolina Department of Natural Resources and Community Development.

ICF Inc. 1988. Pollution Prevention Benefits Manual. First Draft, December 6, 1988. Prepared for U.S. Environmental Protection Agency, Office of Solid Wastes and Office of Policy, Planning, and Evaluation. Fairfax, Va.: ICF Incorporated.

Kneese, A.V., and B.T. Bower. 1968. Managing Water Quality: Economics, Technology, Institutions. Baltimore, Md.: Johns Hopkins University Press. 328 pp.

Kneese, A.V., and B.T. Bower. 1979. Environmental Quality and Residuals Management: A Report of a Research Program on Economic, Technological, and Institutional Aspects. Baltimore, Md.: Johns Hopkins University Press. 337 pp.

Linnerooth, J., and A.V. Kneese. 1989. Hazardous waste management: A West

German approach. Resources 96:7-10.

McHugh, R. 1989. Economic Incentive Mechanisms: An Overview. Draft. Washington, D.C.: Regulatory Innovations Staff, U.S. Environmental Protection Agency. 11 pp.

Melosi, M.V. 1981. Garbage in the Cities: Refuse, Reform, and the Environment, 1880-1980. College Station: Texas A&M University Press. 268 pp.

Miller, D. 1987. Strategy making and structure: Analysis and implications for performance. Acad. Manage. J. 30:7-32.

Muir, J.M., W.R. Muir, and H.V. Sheevers. 1989. Hazardous Waste Reduction Review: The Pretreatment Program. Project No. P50579. Trenton: Division of Science and Research, New Jersey Department of Environmental Protection.

Naj, A.K. 1988. See no evil: Can $100 billion have "No Material Effect" on balance sheets? Wall St. J. 211(May 11):1.

National Toxics Campaign. 1989. Toxics Use Reduction: From Pollution Control to Pollution Prevention. Policy paper, February 1989. Boston: The National Toxics Campaign. 10 pp.

NRC (National Research Council). 1983. Management of Hazardous Industrial Wastes: Research and Development Needs. National Materials Advisory Board Pub. No. NMAB-398. Washington, D.C.: National Academy Press. 85 pp.

NRC (National Research Council). 1985. Reducing Hazardous Waste Generation: An Evaluation and Call for Action. Washington, D.C.: National Academy Press. 76 pp.

NRC (National Research Council). 1986. Ecological Knowledge and Environmental Problem-Solving: Concepts and Case Studies. Washington, D.C.: National Academy Press. 388 pp.

NRC (National Research Council). 1987. Multimedia Approaches to Pollution Control: A Symposium Proceedings. Washington, D.C.: National Academy Press. 155 pp.

NRC (National Research Council). 1989. Alternative Agriculture. Washington, D.C.: National Academy Press. 448 pp.

NRC (National Research Council). 1990. Tracking Toxic Substances at Manufacturing Facilities: Engineering Mass Balance Versus Materials Accounting. Washington, D.C.: National Academy Press.

OTA (U.S. Congress. Office of Technology Assessment). 1986. Serious Reduction of Hazardous Waste: For Pollution Prevention and Industrial Efficiency. OTA-ITE-318. Washington, D.C.: Office of Technology Assessment, U.S. Congress. 254 pp.

OTA (U.S. Congress. Office of Technology Assessment). 1987. From Pollution to Prevention: A Progress Report on Waste Reduction. OTA-ITE-347. Washington, D.C.: Office of Technology Assessment, U.S. Congress. 55 pp.

OTA (U.S. Congress. Office of Technology Assessment). 1989. Facing America's Trash: What's Next for Municipal Solid Waste? OTA-O-424. Washington, D.C.: U.S. Government Printing Office.

Porter, M.E., ed. 1986. Competition in Global Industries. Cambridge, Mass.: Harvard Business School Press. 581 pp.

Prete, P.J., M.B. Edelman, and R.N.L. Andrews. 1988. Solid Waste Reduction: Alternatives for North Carolina. Raleigh: Pollution Prevention Pays Program, North Carolina Department of Natural Resources and Community Development.

Royston, M.G. 1979. Pollution Prevention Pays. London: Pergamon. 197 pp.

Sarokin, D.J., W.R. Muir, C.G. Miller, and S.R. Sperber. 1985. Cutting Chemical Wastes: What 29 Organic Chemical Plants are Doing to Reduce Hazardous Wastes. New York: INFORM. 535 pp.

Schecter, R.N. 1989. Summary of State Waste Reduction Programs. Raleigh, N.C.: National Roundtable of State Waste Reduction Programs. 16 pp.

Scott, W. 1987. Organizations: Rational, Natural, and Open Systems. Englewood Cliffs, N.J.: Prentice-Hall. 377 pp.

Sherry, S. 1988a. Low Cost Ways to Promote Hazardous Waste Minimization: A Resource Guide for Local Governments. Sacramento, Calif.: Local Government Commission.

Sherry, S. 1988b. Minimizing Hazardous Wastes: Regulatory Options for Local Governments. Sacramento, Calif.: Local Government Commission.

Sherry, S. 1988c. Reducing Industrial Toxic Wastes and Discharges: The Role of POTWs. Sacramento, Calif.: Local Government Commission.

Stavins, R. 1988. Project 88: Harnessing Market Forces To Protect Our Environment. Sponsored by Senators Timothy Wirth and John Heinz, December 1988. Washington, D.C.: U.S. Congress.

Todd, R. 1989. Accounting and the Environment: Patching the Information Fabric. Paper prepared for the National Research Council Workshop on Waste Reduction Research Needs, Annapolis, Maryland, May 8-9, 1989 (Board on Environmental Studies and Toxicology, Committee on Opportunities in Applied Environmental Research and Development). 23 pp.

Van de Ven, A.H., H.L. Angle, and M.S. Pool, eds. 1989. Research on the Management of Innovation. Cambridge, Mass.: Ballinger.

Winkler, R.C., and R.A. Winett. 1982. Behavioral interventions in resource conservation: A systems approach based on behavioral economics. Am. Psychol. 37:421—435.

Wolf, K. 1988. Source reduction and the waste management hierarchy. J. Air Pollut. Control Assoc. 38:681-686.

Wolf, K. 1989. Source Reduction: What Is It and How Can We Accomplish It? Paper prepared for the National Research Council Workshop on Waste Reduction Research Needs, Annapolis, Maryland, May 8-9, 1989 (Board on Environmental Studies and Toxicology, Committee on Opportunities in Applied Environmental Research and Development). 31 pp.

Workshop Participants and Agenda

Research Needs For Waste Reduction
Annapolis, Maryland
May 8 & 9, 1989

Richard N.L. Andrews, Chair, University of North Carolina, Chapel Hill
Nicholas Ashford, Massachusetts Institute of Technology, Cambridge
Blair T. Bower, The Conservation Foundation, Arlington
T. Randall Curlee, Oak Ridge National Laboratory, Oak Ridge
Harry Fatkin, Polaroid Corporation, Cambridge
Jeanne Herb, New Jersey EPA, Trenton
Gregory Hollod, CONOCO, Inc., Houston
Hank Garie, New Jersey EPA, Trenton
Joanne Linnerooth, Resources for the Future, Inc., Washington, D.C.
Albert L. Nichols, Harvard University, Cambridge
Richard N. Osborn, Wayne State University, Detroit
Michael Overcash, North Carolina State University, Raleigh
Roger Schecter, North Carolina Department of Natural Resources & Community Development, Raleigh
Rebecca Todd, New York University, New York
Katy Wolf, Source Reduction Research Partnership, Los Angeles

Environmental Protection Agency

David Berg, TIEC
Carl Gerber, ORD
Gerald Kotas, OPPE
Fred Lindsey, ORD
Ron McHugh, Regulatory Innovations Staff

Workshop Agenda

**Committee on Opportunities in
Applied Environmental Research and Development
"Research Needs For Waste Reduction"**

Annapolis, Maryland

May 8 & 9, 1989

Monday, May 8, 1989

8:00 a.m.	Continental Breakfast	Crown & Crab Room
8:30	Plenary Session Welcome and Introductions; background, purposes and proposed schedule	Duke of Gloucester Room Pete Andrews
	Brief perspectives on issues and research approaches:	Katy Wolf Greg Hollod Blair Bower Rebecca Todd Nick Ashford
	Discussion	
10:30	Break	
10:45	Working Groups - see attached list for individual assignments	
12:00 p.m.	Lunch	King of France Tavern
1:30	Working Groups	
3:30	Break	
3:45	Plenary Session - interim report from each working group; discussion	Duke of Gloucester Room
5:00	Adjourn	
6:00	Reception	Atrium
6:30	Dinner	East Chamber

Tuesday, May 9, 1989

8:30 a.m.	Continental Breakfast	Crown & Crab Room
9:00	Reconvene Plenary Session	Duke of Gloucester Room
9:30	Working Groups (set your own break time)	
12:00 p.m.	Lunch	King of France Tavern
1:30	Plenary Session	Duke of Gloucester Room
3:00	Adjourn	

Workshop Papers

RESEARCH NEEDS FOR WASTE REDUCTION *

Richard N. L. Andrews

The purpose of this workshop is to provide recommendations on research needs and opportunities in the field of waste reduction. Recognizing the substantial numbers of similar workshops and reports that have already been devoted to engineering and industrial process research needs in this field, it is intended that this workshop focus particularly on research questions involving three other domains:

(1) the definition and measurement of waste reduction; (2) institutional and behavioral barriers; and (3) policy incentives for waste reduction.

All these domains are important, transcend particular industrial processes, are unlikely to be addressed adequately by the private sector alone, and may therefore be strong candidates for research attention by EPA and other agencies. To address them, invited participants include not only experts on waste reduction per se, but also scholars in related fields who may be able to add insights from their own disciplines to the discussion.

As a working agenda, I propose that we spend the initial session in plenary discussion of the basic issues of waste reduction that deserve our attention, then break into three working groups, each addressing research needs in one of the three domains listed. We will return to plenary sessions for interim reports and discussion at the end of the first day and at least twice during the second; additional suggestions, either by individuals or by any informal groups that may emerge, will also be welcome.

The charge to each group in to identify as specifically as possible the questions related to their topic that deserve research attention, and why; to suggest methods by which they might usefully be approached; and insofar as possible; to recommend priorities among them.

This paper is offered as an initial framework for organizing discussion of the subject of waste reduction. The list of topics is undoubtedly not exhaustive; the framework is itself tentative and open for discussion. Comments, additions and refinements are welcome.

Background

A root cause of most environmental pollution problems is the emission or discard of waste materials

---

*Background paper prepared for the NRC Workshop on Waste Reduction Research Needs, Annapolis, MD, May 8-9, 1989.

and energy (in economic terms, residuals) from human production and consumption activities into the air, water, and land. Policies to reduce such pollution focussed initially on "safe" disposal (increased dilution in air and water, "sanitary landfills" instead of open burning dumps), and subsequently on waste "treatment" technologies, which changed the physical or chemical form of the waste materials before discharging them to the environment. The effect of such policies was sometimes to reduce the quantity or toxicity of some waste streams, but often simply to displace them to other places, future times, or other environmental media. These problems were well characterized by researchers in the late 1960s, but not widely popularized until a decade or more later (cf. Kneese and Bower, 1979; Royston, 1979).

As early as 1976, the U.S. Environmental Protection Agency proposed a hierarchy of waste management options in which waste reduction (as opposed to treatment or disposal) was identified as the preferred approach (41 Fed. Reg. 35350); and the Hazardous and Solid Waste Act of 1984 contained an explicit policy directive "that wherever feasible, the generation of hazardous waste is to be reduced or eliminated as expeditiously as possible." EPA's Office of Research and Development in 1986 authored a "waste minimization strategy," which advocated waste reduction but defined it broadly to include subsequent treatment, storage and disposal as well as reduction at the source; it also was limited to hazardous wastes, and explicitly addressed only technical rather than regulatory and economic barriers (USEPA, 1986).

EPA's Science Advisory Board, in its review of the 1986 strategy document, urged that EPA take a broader view of waste minimization, not limited either to hazardous wastes or even to substances traditionally viewed as "wastes:" it should include any non-product substance that leaves a production process or a site of product handling or use." The SAB also urged special emphasis on "waste prevention (source reduction)," which it defined as a subset of waste minimization practices that focus on in-process practices, as well as on waste generation practices by product users and consumers, that prevent or reduce waste generation per se (USEPA, 1987).

In 1987-88, the SAB sponsored a committee study on environmental research strategies for the 1990s, and the report of this committee (the "Alm Committee") included major emphasis on research needs for "risk reduction" (USEPA, 1988a, 1988b). This report urged that risk reduction be adopted as the central goal both of EPA generally and of its research and development activities; reaffirmed waste prevention (source reduction) as the preferred strategy for risk reduction; and urged that EPA develop a strong program of research related to questions in these areas that were unlikely to be undertaken by or duplicative of research by the private sector. The report noted explicitly that such research should include both technology-based strategies and other strategies involving disciplines other than the physical and biological sciences and engineering; examples of the latter included (among others) policy and economic incentives for risk reduction, risk communication and perception, environmental management and control systems, and education and training programs.

Most recently, in January 1989 EPA issued a formal policy statement adopting pollution prevention through source reduction as an agency-wide policy goal (replacing the narrower focus on 'waste minimization" for legally defined 'hazardous wastes"), and establishing a new Office of Pollution Prevention (in the Office of Policy, Planning, and Evaluation) to develop and implement this purpose across all EPA programs and all three environmental media (USEPA, 1989a). Key components of this program are to include the creation of incentives and elimination of barriers to pollution prevention, efforts at cultural change emphasizing the opportunities and benefits of pollution prevention, and related research and educational activities.

A Pollution Prevention Research Plan is also in preparation by EPA's Office of Research and Development, but as of its February 1989 draft this plan continued to give primary emphasis to process research and technology transfer, product research,, and recycling and rouse; initiation of non-technological research and innovative research of other sorts was given lower priority (USEPA, 1989b).

It appears clear therefore that the deliberations of this workshop may assist EPA (as well as perhaps other federal and state agencies) in defining important topics of research in support of its emerging policy directions, and that this need is unlikely to be met by existing research planning within EPA's Office of Research and Development.

Previous NRC Studies

Study committees of the National Research Council have addressed related topics in several past reports. A 1983 report on management of hazardous industrial wastes endorsed the hierarchical preference for source reduction, and noted that satisfactory management of hazardous industrial wastes is inhibited by nontechnical as well as technical factors, but it did not offer research recommendations on these topics (NRC, 1983). A 1985 report was explicitly directed to institutional factors in reducing waste generation, but it was limited to hazardous wastes, and its recommendations concerning nontechnical factors focussed mainly on implementation and educational programs; its research recommendations were limited to technological methodologies (NRC, 1985).

Recent or current studies on related topics also include committees on multi-media pollution control and on the use of mass-balance information in environmental management and regulation.

Research Issues in Waste Reduction

Three primary domains of research have been identified as primarily candidates for our discussions: the definition and measurement of waste reduction, institutional and behavioral barriers, and policy incentives. If other important research questions emerge from the discussions, they too are welcome, though hopefully in addition to (not instead of) topics within these domains.

I. DEFINITION AND MEASUREMENT OF WASTE REDUCTION

(Background: see esp. papers by Wolf, Bower, Ayres)

Waste reduction requires definition and measurement at two levels: the "micro" level of the waste generator (business firm or operation, household or institution, urban jurisdiction, etc.), and the "macro" level of the aggregate of human processes of materials and energy extraction, conversion, use, and discard.

At the micro level the focus is on the material and energy flows and associated economic benefits and costs to a particular decision unit. This level (and more specifically, decisions in the industrial sector) has been the primary focus of most discussion of waste reduction to date, with less attention as yet to waste reduction either in agriculture and other sectors or in society as a whole.

From the macro level perspective of public policy, however, all materials and much energy become "wastes" at some points in these processes; and conversely, some deliberate uses of-materials are inherently dispersive (cf. Katy Wolf's METH examples, and pesticides and fertilizers generally). If we limit our attention to the perspectives of particular firms or industries, or to "wastes" as a preconceived category, we are likely both to set faulty priorities and to miscount as waste reduction actions which merely displace potential pollutants from one location or process to another.

It is important, therefore, to identify what research is needed to define and measure waste reduction at both these levels. It may be useful for this group to divide its time into several periods, devoted respectively to definition and measurement of waste reduction at the micro and macro levels; and within the micro level, to possible variations in definition and measurement needed across differing sectors and processes.

This task may require consideration of such questions as the following:

1. Why waste reduction? Waste reduction is now widely advocated, but for varied, often unspecified, and sometimes conflicting reasons: reduction of public health risks, of damage to ecosystems, of natural resource use rates, of disposal costs to local governments, of the need for new disposal sites, or merely of inefficiencies in economic production and consumption processes? What research could assist in defining internally consistent strategies for waste reduction, at both micro and macro levels, and in evaluating their impacts and trade-offs?

2. What waste reduction? The operational objectives of waste reduction are also ambiguous as yet: to reduce waste volumes requiring disposal, toxicity of wastes requiring disposal, overall use of toxic materials ("toxics use reduction"), or overall use of materials and energy? or to achieve some other desired pattern of materials and energy use (and if so, what)? What research would help to more clearly define policy objectives for waste reduction, and appropriate measures of its achievement?

3. What differences in definition and in approaches to waste reduction may be required in different types of decision units (e.g. extraction and agriculture, primary materials processing, secondary mfg. and product formulation, packaging/container producers, recycling/reuse businesses, etc.; large integrated firms vs. small specialized firms; etc.)? Is waste reduction best pursued by targeting specific ("highrisk?") substances throughout their processes of extraction and use (e.g. CFCs, lead, chlorine?); by targeting particular stages of the waste generation process (extraction, manufacturing, commercial use, consumer use, waste management); or by targeting particular sectors, industries, or firms that are either especially wasteful, especially hazardous, or especially attractive for opportunistic, cost-effective waste reduction? What research is needed to provide answers to these questions?

4. How might waste reduction best be measured, at both macro and micro levels, and what research Might assist in defining appropriate measurement systems?

II. INSTITUTIONAL & BEHAVIORAL BARRIERS TO WASTE REDUCTION
(Background: see esp. papers by Hollod, Todd, Ashford)

The slogan most widely used to promote waste reduction has been that "pollution prevention pays," not just in societal terms but in the coin of direct self-interest to the waste generator; and a growing list of anecdotes has been adduced to support this claim (e.g. Royston, 1979; Huisingh et al., 1985; Sarokin et al., 1985; and U.S. EPA, 1987b).

Even when presented with information purporting to show such benefits to self-interest, however, more than a few waste generators show surprisingly little interest in change. There appear to be important institutional and behavioral barriers to waste reduction that are not yet well understood even within the business sector; and there is every reason to believe that similar or additional barriers exist in the behavior of other waste generators: agriculture, nonprofit institutions such as universities and government agencies, households, etc. A second important area of research, therefore, is to identify what these barriers are, which ones are present in the behavior of the various types of waste generators, and how they can be most effectively reduced.

It may be useful for this group to divide its time into several periods, each focussing on research needs on institutional and behavioral barriers to waste reduction in a different sector (e.g. businesses, local governments, households). Answering these questions may require consideration of such questions as the following:

1. Who reduces waste and who doesn't, and why? In the business sector, is interest in waste reduction more characteristic of particular types of firms (e.g. large vs. small, resource extraction vs. basic chemicals vs. diversified consumer product manufacturing firms vs. others)? Or of firms with particular characteristics (accounting practices, leadership commitment, "corporate culture," research/innovation capability, etc.)? Outside the business sector, what differences are evident between those who actively reduce wastes and those who don't (institutions, government units, households, etc.), and why?

2. For each type of decision unit, what are the principal barriers to further reduction? Lack of technologically or economically feasible options, of information about options, of willingness to risk new innovations? Inconsistency with other objectives (short term and long term profits, cost minimization, product characteristics and marketing strategies, convenience packaging, etc.)? Or more basic patterns of personal preference and organizational norms, perceived self interest, cultural values, etc.? What research could assist in identifying these barriers more clearly, and in evaluating interventions intended to reduce them?

3. What role do attitudes and perceptions play (on the part of business managers, workers, senior executives, consumers, etc.) in promoting or retarding waste reduction? To what extent do these attitudes

vary: e.g. by type of business, by size and internal differentiation of firms, and by functional responsibilities within the firm (e.g. product design, manufacturing, sales, marketing, envr. health & safety, etc.); and in the consumer sector, by socioeconomic status and other factors? What incentives and communication strategies would most effectively promote waste reduction behavior on the parts of these varied actors?

4. What other changes in business decision processes -for instance, in organizational design, accounting procedures, and internal incentives -- might contribute to waste reduction, and what research might assist in evaluating these possibilities?

5. What existing bodies of research might be applicable to increasing waste reduction behavior: accounting practices, and other economic incentives? studies of corporate strategic planning and management decisionmaking? organizational psychology and human factors research? agricultural extension research on the acceptance of innovations? consumer and household behavior, and marketing research? others? What specific theories, methodologies, and lines of inquiry might be promising avenues for study?

III. PUBLIC POLICY INCENTIVES FOR WASTE REDUCTION (Background: see esp. papers by Curlee, Nichols, McHugh)

Pollution prevention pays, up to a point, because of unrecognized opportunities for more efficient materials use based on self-interest: there is normally at least some "slack" that can be identified between the idealized model of rational economic behavior, and the institutional and behavioral reality of unexamined assumptions, imperfect accounting practices, standard operating procedures, habits, imperfect knowledge of better practices, "lumpiness" of the capital costs of corrective measures, etc.

Beyond that point, however, pollution prevention pays only because some cost factors have changed; and many of these factors are heavily influenced -- sometimes deliberately, but often inadvertently or even perversely -by public policy measures. Some pollution prevention that did not pay yesterday pays today (e.g. because of rising landfill charges and required control technologies); and some pollution prevention that still does not pay today might pay tomorrow (e.g. because of even more costly disposal charges, regulatory expectations, changes in liability doctrines, etc.). Conversely, some pollution prevention that pays today might not pay tomorrow: e.g. if regulatory enforcement is relaxed (giving competitors a free ride), if new requirements require further or different reconfiguration of the same processes, if markets for recycled materials become glutted, or if today's waste reduction proves to have unanticipated adverse consequences of its own.

A third important area of potential research needs, therefore, concerns the effects and effectiveness of public policy incentives -- state and local as well as national -in promoting or retarding waste reduction. As in Group II, it may be useful to subdivide the group's time into periods devoted to different sectors (businesses, government agencies, and households). This area may require consideration of such questions as the following:

1. What categories of wastes might we most want to reduce, and what policy incentives might be most effective for those categories (packaging materials? toxic chemicals? energy? others? Cf. Group I's charge)?

2. What existing public policies increase and decrease, respectively, the incentives for waste reduction in each sector; and what research might assist in answering this question? (Examples: regulations, and regulatory uncertainty; enforcement practices and expectations; educational and technical assistance services; non-environmental policies influencing business decisionmaking; liability standards, and disclosure requirements; taxes, subsidies, and stabilization measures differentially affecting raw and secondary materials markets; etc.)

3. Could waste reduction be better effected by mechanisms to "charge back" disposal costs to waste generators? or farther back yet, to product designers and producers (e.g. deposit-refund and ticket systems), or extraction industries? What research would assist in evaluating both the effectiveness and the potential side effects of such mechanisms? And more generally, what would be the implications, both economic and environmental, of different allocations of the costs of waste management (of which vast reduction is a part)

among raw material producers and manufacturers (and their investors), consumers, taxpayers, secondary material producers, and future generations?

4. What policy instruments appear to have the most effective positive influence on waste reduction behavior of end-users (individuals, households, retail businesses, institutions, etc.)? What instruments have been ineffective or create perverse incentives?

> (Examples: product design requirements or disposal restrictions, availability of recycling services, deposit-return and ticket systems, waste-end taxes, voluntary or mandatory source separation of recyclables, restrictions or charges on amounts of waste generated, special charges or restrictions on more toxic products/wastes, etc.).

5. What existing bodies of research might be applicable to study of the effectiveness of policy incentives for waste reduction? Are there important approaches that should be considered in addition to work on economic and regulatory incentives? public and private service delivery approaches for waste management? behavioral, educational, and communication approaches? Could waste reduction be "marketed" to consumers, procurement officers, and other product users?

6. What research would best advance our understanding of the economics of waste management, including analysis of the trade-offs among waste reduction, recycling, combustion, composting, landfilling, and other management options -- and from a macro perspective as well as the various micro perspectives of businesses local governments, etc.? what influences do public policy incentives have on these trade-offs, and what new incentives might usefully be evaluated? (Example: should recycling/reuse enterprises be subsidized as "infant industries," either directly or through procurement preferences, until they can compete effectively against raw-materials producers?)

7. Are waste disposal costs currently being accounted for in ways consistent with their true meaning and magnitude? What difference would changes in these practices make as incentives for waste reduction?

> Examples: disposal fees based upon costs of past versus of future facilities; unskilled separation/recycling jobs treated as costs, vs. as benefits if individuals would otherwise be unemployed and require social services.

Caveat

In closing, please note that all these suggestions are intended to stimulate rather than constrain your own contributions; please use them and go beyond them in that spirit! I look forward to a rewarding two days with you.

PARTIAL BIBLIOGRAPHY

Allen, D. W. 1988. Policy and Program Options for Reduction of Hazardous Waste in Texas. Houston, TX: National Toxics Campaign.

Andrews, R.N.L. 1988. Waste Reduction for Toxic Substances. Paper presented at the Toxic Waste Symposium, University of California at Santa Cruz, June 24.

Ayres, R.U. 1988. Industrial Metabolism. Draft book chapter, November 29, 1988 (Dept. of Engineering and Public Policy, Carnegie-Mellon University).

Belzer, R.B. and A.L. Nichols. 1988. Economic Incentives To Encourage Hazardous waste Minimization and Safe Disposal. Prepared for U.S. EPA, Office of Policy, Planning, and Evaluation (Cooperative Agreement No. CR813491-01-2).

Bothen, M. and U-B. Fallenius. 1982. Cadmium: Uses, Occurrence, Stipulations. Solna, Sweden: National Swedish Environmental Protection Board.

Curlee, T.R. 1988. Source Reduction and Recycling as Municipal Solid Waste Management

Options: An Overview of Government Actions. Prepared for U.S. EPA, Office of Policy, Planning, and Evaluation and office of Solid Waste. Oak Ridge, TN: Oak Ridge National Laboratory.

Huisingh, D. et al. 1985. Profits of pollution Prevention: A Compendium of North Carolina Case Studies in Resource Conservation and Waste Reduction. Raleigh, NC: Pollution Prevention Program.

ICF Inc. 1988. Draft Pollution Prevention Benefits Manual. Prepared for U.S. EPA, Office of Solid Wastes and Office of Policy, Planning, and Evaluation, December 1988.

Kneese, A.V. and B.T. Bower. 1979. Environmental Quality and Residuals Management. Baltimore, MD: Johns Hopkins University Press.

Linnerooth-Bayer, J. 1988 (Draft). Hazardous Waste Management in the Federal Republic of Germany: The Bavarian and Hessian Systems. (currently at Resources for the Future Inc., Washington, DC).

McHugh, R. 1989. Economic Incentive Mechanisms: An Overview. Washington, DC: U.S. EPA Regulatory Innovations Staff.

Muir, W.R. and J. Underwood. 1988 Promoting Hazardous Waste Reduction: Six Steps States Can Take. New York: INFORM.

National Research Council. 1983. Management of Hazardous Industrial Wastes: Research and Development Needs. Publication NMAB-398. Washington, DC: National Academy Press.

National Research Council. 1985. Reducing Hazardous Waste Generation: An Evaluation and Call To Action. Washington, DC: National Academy Press.

Prete, P.J.; Edelman, M.B.; and R.N.L. Andrews. 1988. Solid Waste Reduction: Opportunities for North Carolina. Raleigh: North Carolina Pollution Prevention Program.

Royston, M. 1979. Pollution Prevention Pays. London: Pergamon.

Sarokin, D.J.; Muir, W.R.; Miller, C.G.; and Sperber, S.R. 1985. Cutting Chemical Wastes. New York: INFORM.

Schecter, R.N. 1987. Summary of State Waste Reduction Efforts. Paper presented at the Massachusetts 4th Annual Waste Source Reduction Conference, Boston, October 21-22, 1987.

Stavins, R.N. 1988. Project 88:.Harnessing Market Forces To Protect our Environment. Washington, DC: Sponsored by Senators Timothy *Wirth and John Heinz, December 1988.

U.S. Congress. office of Technology Assessment. 1986. Serious Reduction of Hazardous Waste. Report No. OTA-ITE-318.

U.S. Congress Office of Technology Assessment. 1987. From Pollution to Prevention: A Progress Report on Waste Reduction. Report No. OTA-ITE-347.

U.S. Environmental Protection Agency. 1986. Report to Congress: Minimization of Hazardous Waste. Washington, DC: USEPA.

U.S. Environmental Protection Agency. 1987a. Review of the office of Research and Development's Waste Minimization Strategy. Report No.-SAB-EEC-88-004 (Science Advisory Board, October 1987).

U.S. Environmental Protection Agency. 1987b. Waste Minimization: Environmental Quality With Economic Benefits. Report No. EPA/530-SW-87-025 (Office of Solid Waste and Emergency Response, October 1987).

U.S. Environmental Protection Agency. 1988a. Future Risk: Research Strategies for the 1990s. Report No. SAB-EC-88-040 (Science Advisory Board, September 1988).

U.S. Environmental Protection Agency. 1988b. Strategies for Risk Reduction Research. Report No. SAB-EC-040E (Science Advisory Board, September 1988, Appendix E).

U.S. Envirormental Protection Agency. 1988c. The Solid Waste Dilemma: An Agenda for Action. Report No. EPA/530-SW-88-052 (Office of Solid Waste, September 1988).

U.S. Environmental Protection Agency. 1989a. Pollution Prevention Policy Statement. Federal Register 54/16:3845-47 (January 26, 1989).

U.S. Environmental Protection Agency. 1989b. Draft Pollution Prevention Research Plan Report to the Congress. (February 15, 1989).

ECONOMIC, ENGINEERING, AND POLICY OPTIONS**
FOR WASTE REDUCTION

Blair T. Bower

INTRODUCTION: SOME BACKGROUND COMMENTS AND QUESTIONS

The issue of options for waste reduction cannot be addressed without first providing some operational definitions and a context for the discussion.

"Waste"---From the view of an activity, a waste is a nonproduct stream of material or energy for which the cost of recovery, collection, and transport for input to another use is greater than the value as an input. This definition is reflected in the diagram in Figure 1.

Sources of wastes in society are indicated in Figure 2.

What types of wastes comprise the focus or foci: liquid, solid, gaseous, energy? all? Wastes of particular concern, e.g., PCBs, dioxin? A partial list of residuals with which society must contend is shown in Table 1. Clearly some of these wastes are going to increase regardless of governmental policies and programs, simply because of increasing population, and corresponding levels of economic activities. Examples include: septage, which will increase as more septic tanks are installed, given that improvement in technology is unlikely; mining residuals, which inevitably increase as the quality of ores decreases and the depth to oil and gas reserves increases#, i.e., more materials and energy input--and hence more residuals generated--p,or unit of output; more construction debris as infrastructure facilities and housing stock need to be replaced; BOD5 and TSS per capita not likely to change, unless bioengineering quickly changes the metabolism of the human species; with finite assimilative capacity of the nation's water bodies--which capacity may decrease for the country as a whole if global warming occurs--larger amounts of sludge from treating liquid residuals from residences will be generated to maintain the same loads on water bodies as at present, which in turn means more materials and energy inputs both to produce the sludge in waste treatment and to dispose of the sludge.

**Paper prepared for National Academy of Science's Committee on Opportunities in Applied Environmental Research and Development Workshop, May 8-9, 1989, Annapolis, MD

Table 1. Partial List of Residuals Generated in Society, Excluding Spills

Mining: overburden, tailings, leachate, surface runoff, wind entrained particles

Agricultural operations: crop: sediment, pesticides, and nutrients in runoff, nitrogenous material and pesticides in leachate to ground water, wind-blown sediment, wind transported pesticides, volatilized pesticides, pesticide and fertilizer containers

Silvicultural operations: suspended sediment and pesticides in surface runoff, wind transported pesticides, slash

Transportation: CO, HC, NO_x, particulates, particles from tires deposited on ground surface, oil, liquids discharged from boats, salt/sand from snow "removal", tires, used oil, batteries, obsolete vehicles

Residential: white goods, bulky materials, septage, food wastes, yard wastes, Aluminum (cans, wrap), steel cans, glass, plastic (containers, wrap, trays, utensils), household batteries, UN, UCC, UMOP, junk mail, liquid residuals, CO, NO_x, SO_2, TSP, ash (depending on fuel and heating system)

Energy generation: CO, NO_x, SO_2, TSP, Ash (fly, bottom, scrubber sludge), water treatment chemicals, suspended solids, boiler and cooling system blowdown

Manufacturing: "priority pollutants" (liquid), BOD_5, TSS, Food wastes, yard wastes, metals, wood, UCC, UN, UMOP, glass, plastic, Aluminum cans, various solid wastes peculiar to manufacturing process/product combinations, e.g., carpet trimmings, cuttings from shooting in manufacture of mobile homes, gaseous residuals from heating/air-conditioning/process steam/electricity generation, water treatment sludge, wastewater treatment sludge, scrubber sludge

Commercial: food wastes, yard wastes, glass, plastic, Al cans, UCC, UMOP, UN, bulky goods, gaseous residuals from heating/air conditioning, liquid residuals such as BOD_5, TSS

institutional: same as commercial plus those peculiar to activities, e.g., infectious wastes from hospitals/nursing homes, low level radioactive materials from hospitals/ research labs

Municipal: street sweepings, sediment from debris basins, dead animals, grass clippings/brush and tree trimming from public parks/streets/public building areas, water treatment sludge, wastewater treatment sludge, septage

Construction wood (lumber + tree segments), bricks, dirt, stamps, fixtures, metals, plastic, food wastes and food containers

From whose view/from what stance is waste reduction to be measured, and how is it to be measured? individual activity? metropolitan area? state? conterminous United States? Does waste reduction occur only if wastes generation is reduced "across the board"? Does waste reduction occur when generation and discharge of certain undesirable wastes are reduced, but discharges of some less undesirable wastes increase?

Many localities and many states are facing the exhaustion of "acceptable" landfill capacity, so that their focus is on disposal of solid wastes. The Director of Public Works of City 0, or the head of the Solid Wastes Division, faces the problem of how to collect, transport, and dispose of X thousand tons per day, and regulate the disposal of an additional Y thousands tons per day. In the case of New York City, this amount, X + Y, is about 26,000 tons per day. Waste reduction, in terms of the director or division head, means

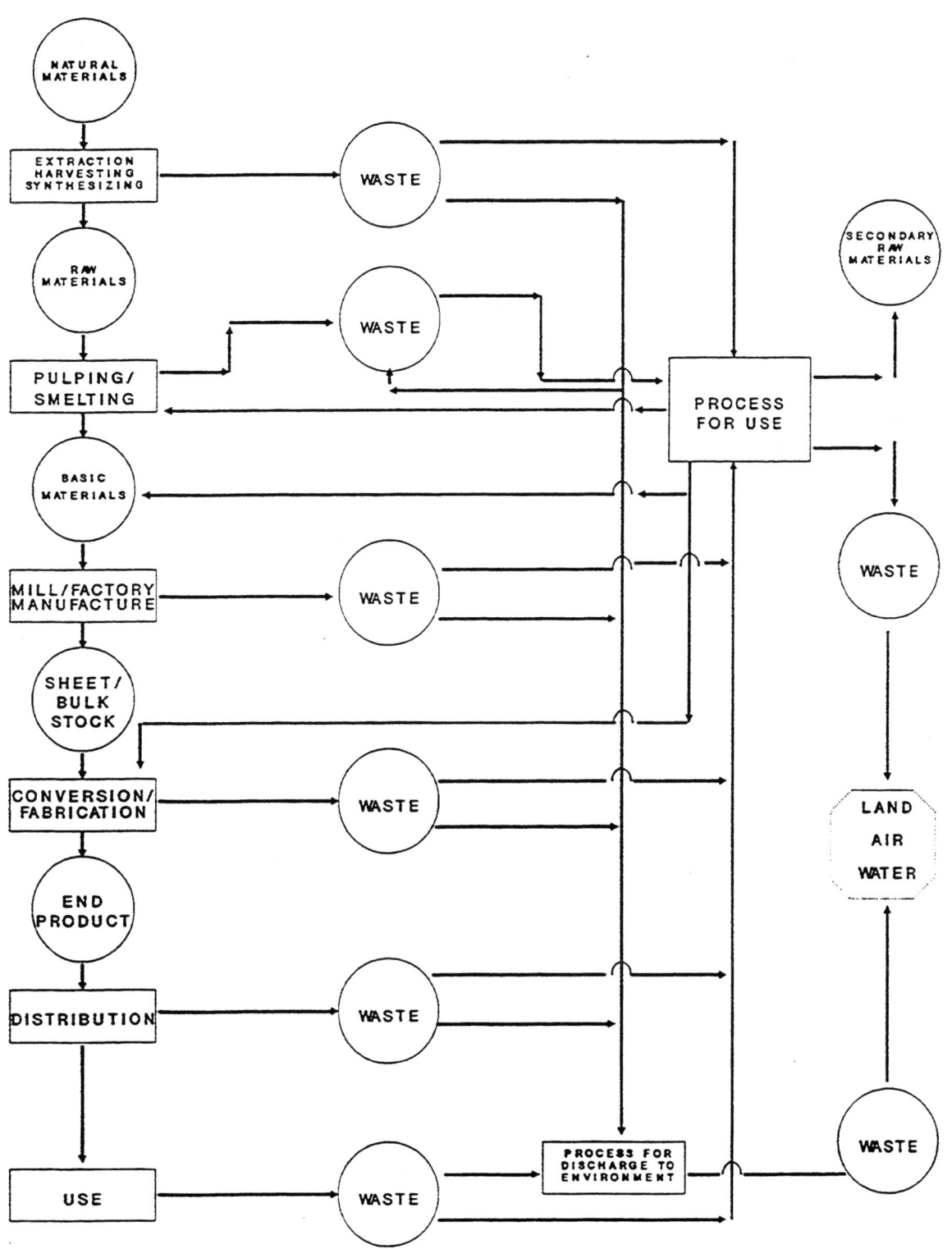

FIGURE 1 Sources of Wastes in Society

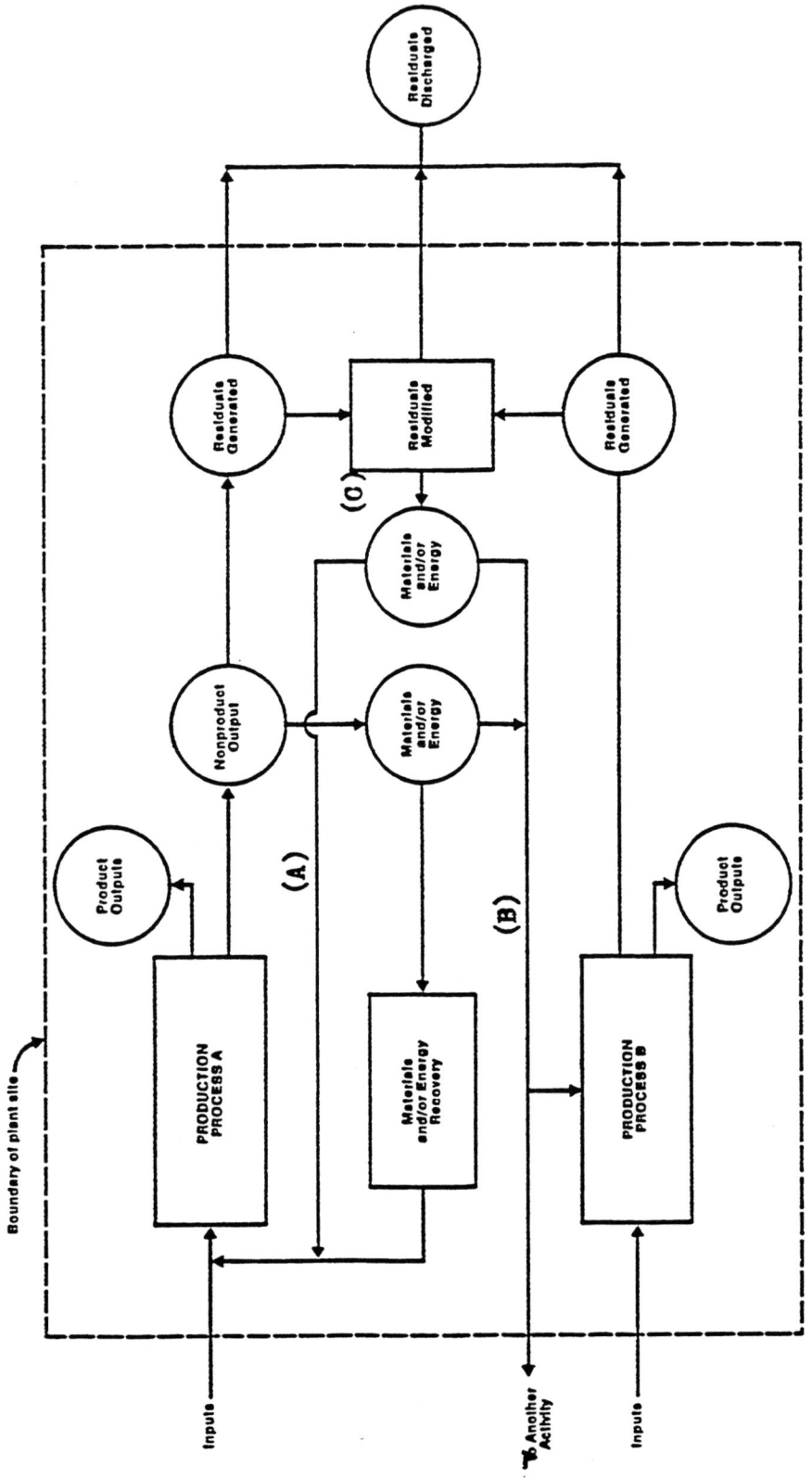

Notes: (A)—Materials and energy recovery will take place up to the level where marginal cost of recovery equals the value of the recovered material or energy, within the context of whatever external constraints exist.
(B)—By-product production path, either in on-site or off-site facilities.
(C)—Residuals modification for materials or energy recovery will take place only to the extent that overall costs of meeting discharge constraints are decreased by such recovery.

FIGURE 2 Definition of Residuals Generation and Discharge

reducing the net amount for which he is responsible in terms of input to the landfill. As illustrated in Figure 3, waste reduction in this view means reducing [$\Sigma \ SR_{Ai} + \Sigma \ SR_{Bj}$]. Reclamation, resource recovery, composing operations generate solid residuals (and other residuals) which require disposal. To the extent that resource recovery and reclamation facilities are outside of his area, the director or solid wastes division head does not have to dispose of the residuals generated in these operations; they have been "exported" to other regions. However, society as a whole must still dispose of them. The balance sheet of the nation may or may not be improved when the net solid waste for disposal in the region has decreased.

Waste reduction could be defined, for society as a whole, to mean reducing:

$$\sum_{g=1}^{T} GG_g,$$

given the nature of materials and energy flows as shown in Figure 4. Σ GG would be the total of liquid, gaseous, solid, and energy residuals, undifferentiated with respect to nature of their environmental effects when discharged. (Theoretically---and only theoretically---weights according to relative toxicity of discharges of materials to different environmental media could be assigned to different materials and energy waste streams.)

The assumption is made in the following that the primary focus of "waste reduction" is solid wastes.

FACTORS INCREASING WASTES GENERATION

Identifying options for waste reduction must be cognizant of the factors which are increasing, and tend to increase, wastes generation per capita in the "affluent, effluent" society, particularly pressures from sales departments and marketing specialists. When Vance Packard wrote The Hidden Persuaders, it was early in the evolution of the disposable product, proliferating products society. Any sequel to that classic written now would indicate that society has gone much further down that road. The ingenuity and imagination of sales/market types are the driving forces. These forces are reflected in: (1) product mix in a given plant and product proliferation; (2) product specifications; and (3) new products. Cursory comments on these follow.

Product Mix and Product Proliferation

Product mix refers to the mix of products and/or services produced by a given activity. For example, a petroleum refinery may produce several grades of gasoline, jet fuel, heating oil, asphalt, lubricating oils; or it may produce only lubricating oils and asphalt. An integrated pulp and paper mill may produce a variety of types and colors of consumer products--napkins, tissues, towels -- as well as printing papers and business forms. The consumer products may be pink, yellow, blue; with or without designs; scented or unscented. Or, an integrated mill may produce only linerboard, but produce from two or three to a dozen grades of linerboard, from 15 pound to 90 pound. A cannery may produce six different styles of canned peaches, the whole range of tomato products, pet food, and soft drinks. Or the cannery may produce only three styles of green beans. An automobile assembly plant may produce 30 models of a given car, with a virtually infinite number of combinations of colors and accessories possible for each model. One can now purchase Green Giant corn niblets in a Microwave One serving, a serving size perfect for one person", as the package cover says, or in a #10 can, and every size in between. The packaging per ounce for the one Serving is several times that for the #10 can, or for the two pound package of frozen corn niblets. Coffee in the supermarket no longer comes in simply a 16 ounce can. Beginning in 1988, take your pick: 16, 13, 11.5, 10. (They all look virtually the same).

In a few cases, increasing the range of the product mix may have positive impacts, in terms of reducing generation of residuals per unit of output. An example is the tomato cannery where most of the

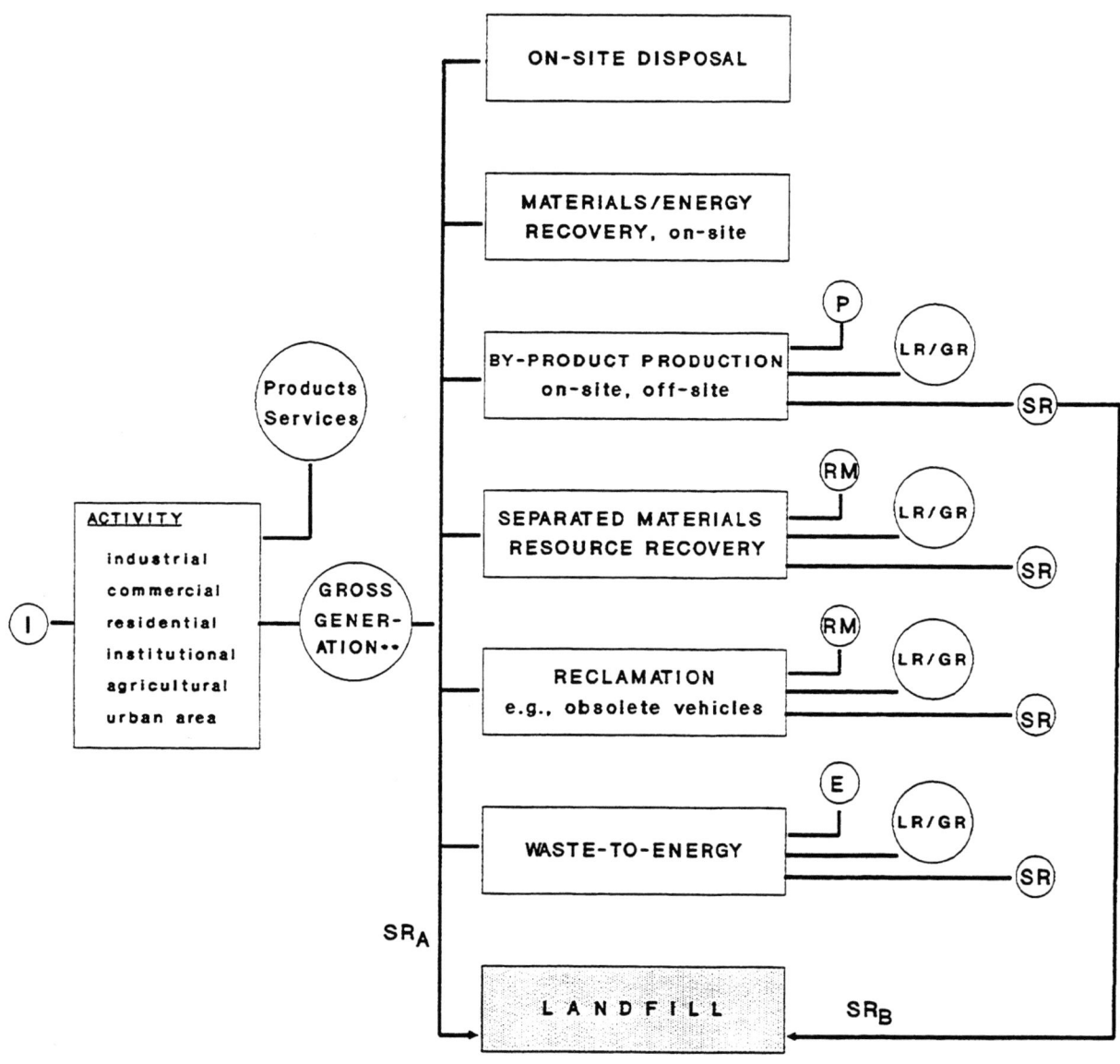

FIGURE 3 Disposition of Solid Residuals in Society

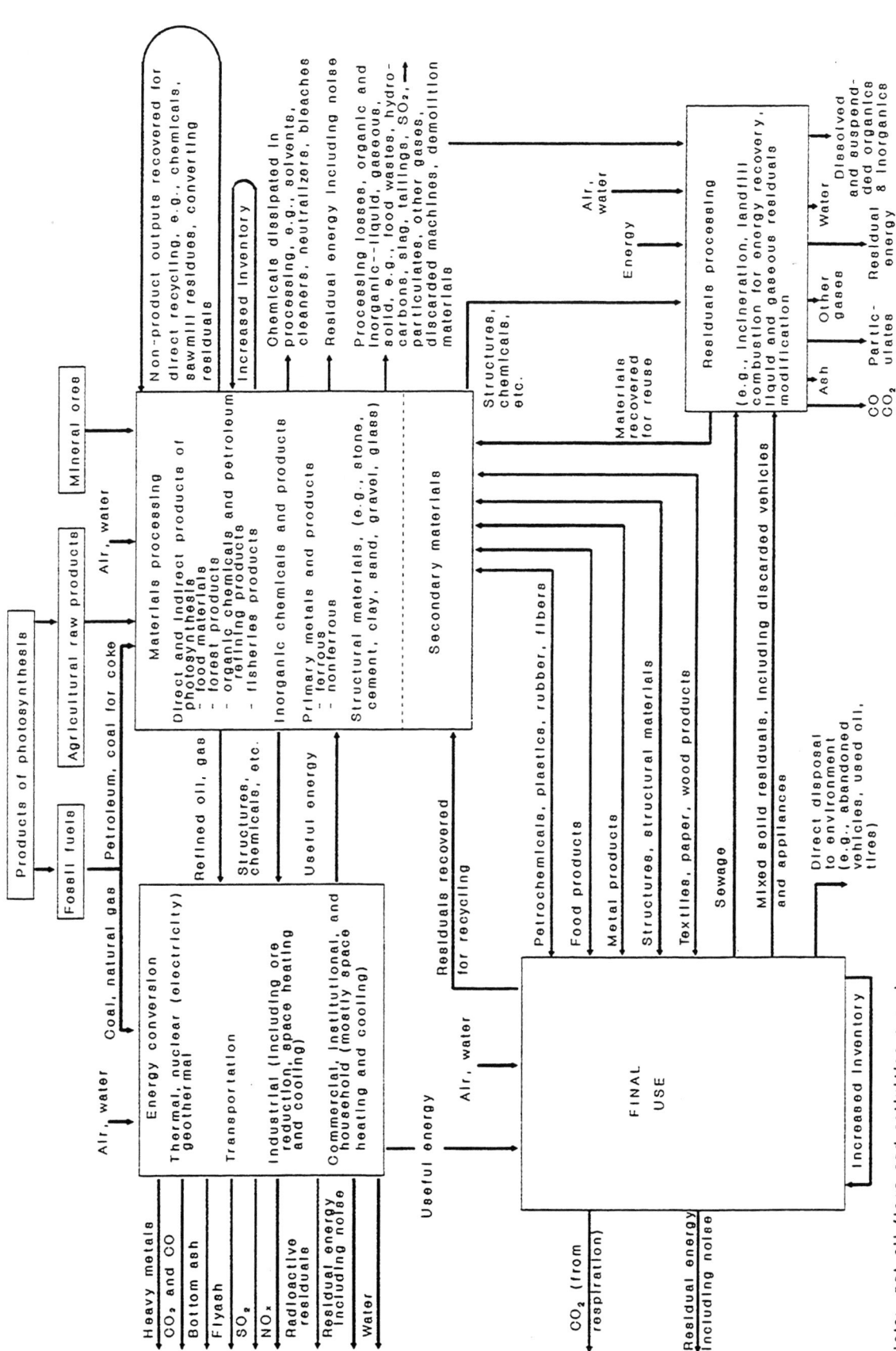

FIGURE 4 Materials and Energy Flows in a Society

tomato, except field dirt and bruised sections, goes into some tomato product, and the remaining pulp is used in the production of pot food.

However, in most cases increasing the range of Products produced in a single plant complicates the production process and results in increased residuals generation per unit of product. The cannery producing products in container sizes from cocktail to #10 requires more inputs and generates more residuals than the cannery processing the same amount of raw material but producing products in containers of only half as many different sizes. To "squeeze" more and more gasoline from a barrel of crude petroleum requires more and more processing, which in turn means more material and energy inputs and more residuals generation per barrel of crude processed, for a given level of technology. (A caveat: In a multi-product chemical plant, where outputs of some processes are inputs to other processes, residuals generation per unit of feedstock may be less than for chemical plants where fewer final products are produced.)

Even the increased efficiency in production resulting from application of digital control in continuous process industries can be diminished by a wider product mix. For example, on modern Fourdrinier paper machines, the standard procedure increasingly is to make grade changes without stopping the machine. This means that for some period of time all of the output is wasted to the broke system, to be returned subsequently to the paper machine, and all water used provides no product output. For example, with a product output of 480 tons per day, or 20 tons per hour, a 15-minute grade change would involve on the order of 5 tons of broke. At 6% moisture off the machine, the amount of water to dilute the paper to one quarter of one percent consistency for direct reuse is about 500,000 gallons. This represents a significant increase in total daily water demand and wastewater generated compared with a production procedure which would shut down between grade changes. Of course, the shut down procedure involves other costs. In general, the more changes in type of product output, e.g., every time the assembly line has to shift--in the refinery, the pap-or mill, the cannery, the automobile assembly plant---the larger residuals generation is per unit of output or per unit of raw product processed, for a given technology. Of course, this principle also applies to production from year to year. That is, changing product mix each year--to the annual fashion--requires significantly more inputs and generates more residuals per unit than changing the product mix/product specifications only once every four or five years, be the product automobiles or women's shoes.

In addition to product mix at a given plant resulting in more residuals generation per unit of product, the same type of result occurs across plants producing a given type of product. Take your pick among at least forty oat bran cereals on the supermarket shelf. If facing that decision gives you a headache, as Krier (1989) suggests, seek a product to relieve your headaches: aspirin, acetaminophen, ibuprofen; regular or extra strength; gelatin coated, enteric coated, safety coated, toleraid micro-coated.

Product Output Specifications

Product output specifications refer to the specific characteristics desired in a given product or service, as measured by standard tests. For example, wet strength, basis weight, and color (in terms of brightness) are characteristics specified for a paper towel.

Four aspects of product specifications merit emphasis. One, the more stringent and exacting the product specifications, generally the greater the waste generation per unit of product. A simple example is shown in Table 2, which shows the quantities of residuals generated in the production of one ton of <u>white</u> tissue paper, General Electric (GE) brightness 80-82, and the quantities generated in producing one ton of <u>unbleached</u> tissue paper, GE brightness 25, using the same raw material and pulping process, and having all other product characteristics the same. Because a high level of bleaching is required to achieve the higher brightness, and because bleaching is a highly energy intensive process, the higher product specification --- solely for appearance --- results in substantially higher residuals generation per unit of product (and correspondingly, more inputs per unit at the "front" end).

Table 2. Residuals Generated in Producing One Ton[a] of Tissue Paper[b]

	Standard Brightness (GEB 80-82)[c]	Unbleached (GEB 25)
	All values in pounds per ton	
Gaseous		
Chlorine	1.2	0
Chlorine dioxide	0.6	0
Sulfur dioxide[d]	5.6/20.0[e]	5.1/7.0[e]
Hydrogen sulfide and organic sulfide	25.5	23.2
Particulates[d]	57.5/1.0[e]	52.4/0.3[e]
Liquid		
Dissolved inorganic solids	263	22
Dissolved organic solids	244	41
Suspended organic solids	113	107
Suspended inorganic solids	4/5	4.1
Five-day biochemical oxygen demand	147	31
Solid		
Inorganic solids	82.0	73.7
Organic solids	0	0

[a] Output is as air-dry paper (6% moisture), equivalent to 1880 pounds on a bone-dry basis

[b] From softwood, using kraft (sulphate) pulping process, with no constraints on discharges

[c] Using CEHD bleaching sequence

[d] 1% sulfur fuel oil assumed for use to generate heating steam and electrical energy for plant use.

[e] First figure relates to production process, second to fuel combustion

Source: Knots, A.V. and Bower, B.T., 1979, Environmental quality and residuals management, Johns Hopkins Press, Baltimore, Table 2, pp. 65-67.

One expert in the paper industry (Day, 1974) expressed the issue thus:

What customer asked for a dazzling white and bright shoot in the first place? Who knows. And who needs it?

The pursuit of higher whiteness and brightness is expensive, but not expensive just in terms of dollars. It's also costly in terms of materials and energy.

I believe that an honest evaluation of the situation will bring the conclusion that present industry standards for brightness have been developed to the extent that a major portion of chemical is used purely for cosmetic purposes, and that the consumer public could be served as well or better by a return to lower brightness standards, not only in publication papers but in tissues, packaging and many specialties.

If Day were writing in 1989, he would be able to state that "cosmetizing" has increased significantly in the industry. For example, over the 1974-89 period a substantial increase in the proportion of linerboard for corrugated cartons being bleached has occurred. This reflects a major trend in the paper industry, namely, instead of shipping containers being simply shipping containers, 'many customers today require boxes that can double as point-of-sale displays featuring colorful graphics to help sell the product" (Kane, 1989, p. 95). The added colors, which could not be used without bleaching of the linerboard, increase the difficulty and costs of using the discarded shipping containers as raw material in paperboard production.

Not only are some of the brightness specifications unneeded for the function, and perhaps even counterproductive, some performance specifications for paper products may well be excessive in relation to uses for which the paper products are designed. Lowe (1973) suggested that the U.S. specifications for paperboard for shipping electronic appliances, e.g., radios, television sets, were substantially higher than the specifications used by the Japanese. Yet the Japanese appliances seem to arrive in operating condition.

Not only the cosmetics of packaging have increased waste generation per unit of product but also the greater use of packaging per unit of product has evolved, particularly with respect to foods. Between 1963 and 1971 the actual weight of food consumed per capita in the U.S. increased 2.3% while the weight of food packaging increased 33.3% per capita. Between 1958 and 1970, the weight of milk consumed per capita decreased by 23% while the weight of milk containers increased 26% per capita. This occurred even though glass containers for milk almost became extinct during the period. The data for the period since 1971 show essentially a continuation of the trend toward increasing packaging per unit of food product, e.g., shipping containers, food containers, and plastic materials (Economic Research Service, 1988).

One of the most pervasive changes in product specifications has been the shift from multiple use items to single use, "throw away" items. These range from nonreturnable containers for beer, soft drinks, and juices, to throw away dishes and cutlery in various institutions, a.g., cafeterias, and on airplanes, to throw away "linens" in hospitals. For example, on one flight of about 2.5 hours from Minnesota to Washington, D.C., the airplane passenger received a total volume of about 2000 cubic centimeters of nonreturnable containers, utensils, and napkins. Except for the napkins, the basic raw material was plastic from petroleum. Disposable items have become the norm in hospitals, nursing homes, clinics, school cafeterias, and fast food emporia. (Note: To point out the fact that this shift increases residuals generation is not to deny the benefits of use of such items, at least in the first three of the above activities.)

Not only have product specifications become more stringent, but at least some types of products have become more complex. This increasing complexity relates particularly to automobiles and appliances. The "average" automobile of the mid-1950s had about 7500 parts; the average of the mid 1970s had twice that many. Washing machines are produced with 59 different cycles. Increased complexity typically increases residuals generation in production and makes repair more difficult. The latter in turn typically leads to more rapid "discard".

New Products

The epitome of the solid wastes generators is represented by those individuals in sales/marketing whose fertile imaginations produce such items as: disposable videocassettes, which will erase themselves after a limited number of plays; colored plastic wrap, in red, blue, yellow, green, brought to you by Reynolds Metals Co., which wrap "makes the mundane task of wrapping leftovers fun" (Associated Press, 1989); this innovation of the consumer products division is a way to "liven up the kitchen" (ibid.); use of facsimile machines to receive weather satellite maps in order to indicate to trout fisherman on Lake Michigan where the influences of cold water and warm water are, which locations are concentration areas of trout (Anon, 1989).

The above represent a small sample, a microcosm of similar items in the Hammacher-Schlemmer catalogue. But as in some other sectors, the U.S. comes in second to the Japanese with respect to inventiveness along these lines. On 26 April 1989, scent-dispensing telephone booths will debut on Namiki Street in Tokyo. Hourly the booths will blow scent over shoppers passing by, accompanied by the tinkling of chimes. Hiatt (1989, p. A 35) explains the situation:

> For industry here (Japan), inventiveness is a matter of necessity. Japanese homes are so cramped and saturated with television sets and videotape recorders that companies must constantly devise products that consumers never know they needed. And when a product catches on, it can become fashionable with remarkable speed.
>
> Such in the case this year with tiny vibrators, designed to relax the muscles of the back, and with $1,000 toilet bowls that spray warm water and then blow warm air across one's backside--a feature that now graces more than 10 percent of Japan's 'western-style' toilets.

Sony has introduced disposable paper watches. They are made in China and cost less than $4 each. "More than a million of the 'peal-off timepieces' have been sold, in dozens of designs" (ibid., p. A35).

The clear implication of the foregoing is that there are large pressures for, and substantial segments of the economy built and dependent on, the continued proliferation of products, the continued annual changes of styles/fashions, the continued emphasis on "cosmetizing" consumer products. Waste reduction, with respect to both of the definitions stated above, would require on "about face" in this "society-set".

TECHNOLOGIC OPTIONS FOR WASTE REDUCTION

Table 3 shows one classification of physical measures (technological options) for reducing residuals discharges to the environment. Category A measures reduce gross generation; Category B measures reduce not quantity of materials for disposal to landfill, for any given sat of production functions.

Generation of residuals in extraction, processing, manufacturing, service activities is a function of the nature/quantity of inputs, technology of production, and product mix/product specifications. Some examples are given in Table 3; others can be cited. For example, with respect to inputs, shift from high input (intensive) agriculture to low input agriculture; in-field sorting of tomatoes for canning, which significantly reduces residuals generation at the cannery; genetic development of tomatoes more amenable to canning (even though they taste less like tomatoes), which reduces generation in canning; producing wood chips for papermaking in the woods rather than at the mill.

There are two basic approaches to technological options with respect to production technology- one involves modification or changes of unit processes or unit operations. The procedure begins with analysis of materials balances and energy balances of each process/operation. The complexity of this analysis is suggested by Figure 5, a simplified flow diagram of pulp and paper production. Alternative unit processes/operations or sets can then be posited and evaluated, such as changing from wet debarking to dry debarking in pulp and paper production; from wet peeling to dry peeling in canning; from ingot casting to continuous casting in steel production; from batch cooking to continuous hydrosterilizers in canning.

The other approach involves improved control over the production process itself, typically through automatic, digital control systems. Adoption of such systems in the chemical processing industries, including pulp and paper, has resulted in significant increases in yields and associated reductions in residuals generation.

If changing product specifications is considered a technological option, the effects of such changes have been suggested in the previous section.

With respect to Category B measures, technological options for reducing the net quantity of solid wastes to landfill are of three major types: technologies for in-plant materials and energy recovery; technologies for reclaiming or recovering useful materials from residuals; and technologies for producing useful products from residuals. The first is exemplified by the relatively simple but effective condensate management system in pulp and paper mills (Hahn, 1989). The second is exemplified by the evolution of shredding-separation-compaction technology combinations to produce steel scrap from obsolete vehicles.

The third is exemplified by recent technological developments in use of residuals for raw materials. Wastewater treatment plant sludge has been used to produce bricks for construction (Bryan, 1984). The first U.S. plant to use discarded polystyrene foam products, e.g., food trays, cups, containers, as raw materials went on line in January 1989 (Anon., 1989b). Several plants have been, and are using polyethylene terephthalate (PET) and high-density polyethylene (HDPE) as raw materials. However, the products produced thus far in the use of PET and HDPE are relatively low grade plastic products. Dow Chemical's now process, although more expensive, reportedly will yield higher quality plastic which can be used to manufacture plastic bottles (Anon, 1988). It is important to note that DOW's process uses trichloroethane, which has been "indicted" as one of the substances damaging the ozone layer. This is an example of the care which must be taken in analyzing technologic tradeoffs, i.e., is the "not" change to society really positive. Another example of the third is the development of shredding technology and the associated combustion technology for using tires for fuel to generate energy. Shredders which reduce stumps to wood chips for use as mulch is another technologic development which is reducing discharges to landfills.

Product output specifications are important not only with respect to gross generation but also with respect to reducing amounts to landfill by increasing use of residuals in production. For the tissue paper example in Table 2, if a brightness of GEB 80-82 is desired, it is not possible to use 100% No. 1 mixed waste paper as raw material. If a brightness of 25 is acceptable, corresponding to unbleached kraft pulp as raw material, 100% waste paper can be used (Bower, et al., 1973).

STIMULI TO WASTE REDUCTION

What induces the adoption of measures/technologic options which result in waste reduction, gross and/or net? An arbitrary classification of stimuli/inducers consists of: (1) external factors; (2) internal factors; and (3) governmental actions excluding water intake charges, sawer charges, and energy charges, which are included under external factors.

External Factors

Changes in prices of factor inputs have been important factors in inducing waste reduction for manufacturing, commercial, and service sectors. These include prices of energy, both electrical energy and fuel; wastewater disposal; solid wastes disposal; feedstocks/chemicals; intake water. The sharply increased energy costs in 1973-74 induced responses by segments of the chemical industry, which included reduction in product mixes and lowered product specifications (see Rabire and Winton, 1974, and A.non., 1974). Sawer charges imposed on the basis of strength, i.e., quantities and/or concentrations of specific constituents, have induced substantial reductions in discharges, via some combination of raw material and/or process change and better "housekeeping", as described by Knees& and Bover (1968, pp. 166-170) and Hudson, at al. (1981, Chapter 3). Increased solid wastes disposal costs and increased raw material costs are inducing waste

Table 3. Classification of Physical Measures for Reducing Residuals Discharges

CATEGORY	SUBCATEGORY	EXAMPLES
A. REDUCE RESIDUALS GENERATION	1. Increase longevity of goods	
	2. Change type of raw material inputs	High to low sulfur crude, fuel oil, coal; concentrated vs. raw ore; use of residuals instead of virgin materials, i.e., aluminum cans instead of bauxite
	3. Change production process including mode and motive power of transport	Individual vehicles to mass transit; ICE to ECE; H_2SO_4 to HCl for pickling steel; CEHDED bleaching to oxygen bleaching; ingot casting to continuous casting; less energy intensive process for producing aluminum
	4. Change final demand	a. Change product mix — reduce number of grades or styles of product, i.e., chemicals, linerboard, paperboard, canned peaches; prohibit non-returnable containers;
		b. Change product specifications — reduce brightness of consumer paper products, such as towels; LAS instead of ABS; short-lived, specific pesticides instead of long-lived, general pesticides; high octane to low octane gas
	5. Change timing of activity	Staggered office hours; change schedule of production
	6. In-plant recirculation of water[1]	In beet sugar production, peach canning
B. MODIFY RESIDUALS AFTER GENERATION, IN ON-SITE AND/OR COLLECTIVE FACILITIES	1. Materials recovery (direct recycle)	Chemical and fiber recovery in paper production; catalyst recovery in petroleum refining; recycling of mill scale in steel production
	2. By-product production (indirect recycle)	a. To final products — tomato pulp into pet food; citrus peels into candy; peach pits into charcoal briquettes; wood products residues into pressed logs
		b. To intermediates — obsolete vehicles into steel scrap/steel; used corrugated containers into linerboard; used aluminum cans into aluminum ingots; sulphite waste liquor to yeast
	3. Modification of residuals streams	Combustion of solid residuals to generate energy; incineration; landfill; composting; compression of solid residuals; land spraying of sludge; precipitation; sedimentation; scrubbing; biological oxidation; chemical oxidation
	4. Effluent reuse	a. Direct — sewage plant effluent for cooling water
		b. Indirect — ground water recharge with modified liquid residual

[1] Only in relatively few types of cases does in-plant water recirculation modify/reduce a residual in a liquid residuals stream. Recirculation does reduce the gydraulic load on any materials recovery or residuals modification facility, thereby reducing residuals modification costs.

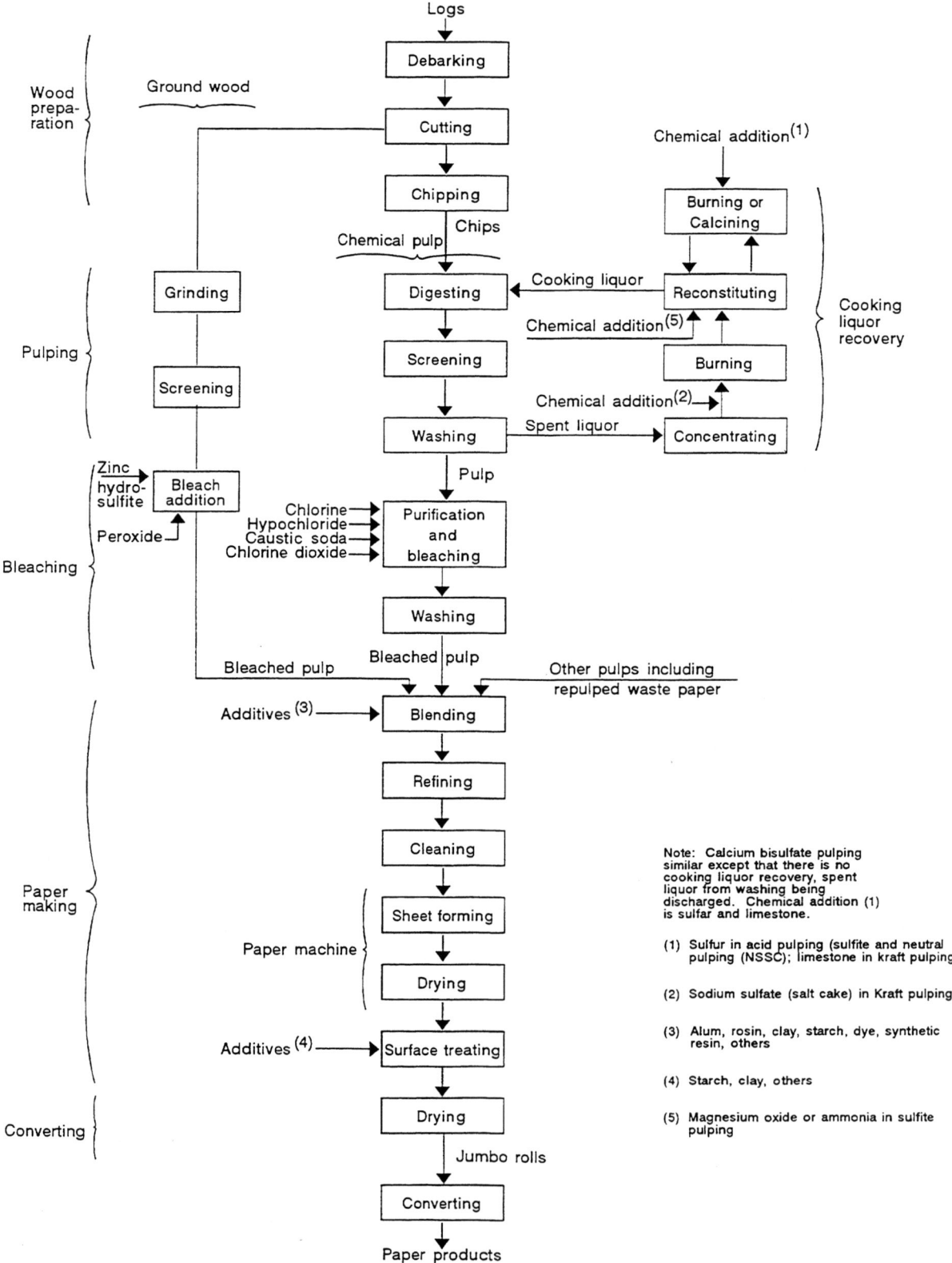

FIGURE 5 Flow Diagram of Pulp and Paper Production

reduction in chemical process industries (Anon., 1987). Similarly, increased solid wastes disposal costs, particularly with respect to hazardous materials, have induced the development of waste reduction "plans" in various states (Dombrowski, 1989).

Other external factors which are inducing waste reduction responses are liability potential, the NIMBY syndrome, and regulations or the "threat" of regulations. The liability potential, i.s., liability for damages resulting from inadequately handling of solid wastes, particularly hazardous wastes (Cheremisinoff, 1989), 'S affective at least in the privata sector. The quits common opposition to having a waste-to-energy facility or a landfill in "my" backyard has, similarly to increased factor input prices, stimulated more effort to reduce generation and to increase resource recovery (Johnson, 1989). The investigation initiated by EPA into dioxin generation in the pulp and paper industry and the announcement that standards with respect to dioxin discharges would be promulgated within about two years has induced various companies to modify production processes to reduce dioxin generation. At least some of these modifications yield an overall net decrease in generation.

Internal Factors

Probably the most effective internal factor, for a private or public entity, is management commitment to and execution of waste reduction/pollution prevention, rather than commitment to reduction in discharge by "end-of-pipe" measures. (The end-of-pipe measures of course actually _increase_ generation and discharge of residuals, albeit different mixes than would occur without the measures.) The Pollution Prevention Pays (3P) program of Minnesota Mining and Manufacturing (3M), initiated in 1975, well exemplifies this approach. Bringer (1988) describes the structure of the program, the results, and the latest step, the inauguration of a five-year Pollution Prevention Plus (3P+) program. This new program, initiated in 1988, will redUCe discharges beyond required levels. However, it is clear that in some or many cases, waste reduction (cum environmental effects) is not a factor considered in design. For example, a 'design evaluation matrix" was developed under the aegis of the Construction Industry Institute (Tucker and Sutherland, 1988). The matrix contains neither a criterion in relation to waste reduction nor a criterion in relation to environmental effects.

Where an activity, public or private, consists of a numb-or of units physically separate, as in a metallurgical plant or university, a system of in-plant charges for water and sawer services can result, and has resulted, in significant waste reduction. Kneese and Bower (1968, pp. 170-172) provide a cursory discussion of such charges.

Other unforeseen factors may lead to waste reduction by activities, either gross or not discharge. For example, initiation in the early 1970S of separation and recovery of various types of office paper in a large bank in San Francisco was stimulated by the interest of the bank president's wife in resource conservation.

Governmental Actions

Governmental entities, at all levels, can intervene at various points in the system with actions which can have positive or negative effects on waste reduction. Figure 6 illustrates possible intervention points.

Governmental actions can take the form of economic incentives; regulations; joint development; and distribution of technical information. Economic incentives include various forms of subsidies in relation to capital costs of waste reduction technologic options in industry, agricultural, commercial, municipal operations, such an grants for a portion of capital costs and below market interest rate loans. Federal and state governments can provide investment tax credits and rapid depreciation allowances for installation of waste reduction technologies, analogous to what has been done with respect to "pollution control" equipment, where such equipment—except in a few cases—has meant end-of-pipe facilities. Increased severance taxes on minerals and elimination of depletion allowances on minerals would increase the costs of virgin materials relative to secondary materials, thereby stimulating additional materials recovery/energy recovery and additional reclamation. Increasing grazing fees to private market levels would reduce stocking and reduce residuals generation. Surcharges on inputs, such as fertilizers and energy, may induce some positive response. Levying or requiring performance bonds for mining operations would have a salutary effect.

Examples of regulations which may induce waste reduction, gross or net, include specification of product ingredients; prohibition of certain products/materials; energy design standards for buildings and appliances; enforcement of discharge limits; enforcement of BMPs; restrictions to encourage use of mass transit; and building design codes for multi-unit commercial buildings, e.g., shopping &alls, and multi-unit residential buildings, which would require designs to enable efficient handling of separated solid residuals.

Technical information development and transfer can begin with jointly financed research projects on methods to reduce wastes generation, via changes-in raw materials, production processes, product specifications, or all three. State PPP programs, as in North Carolina, are mechanisms to provide information. Probably expenditures on such program an order of magnitude larger would still yield positive not benefits.

What of governmental actions in relation to other governmental agencies and programs? Wastes generation problems in agencies under government or quasi-government jurisdiction are legion at many DOE installations, such as Aberdeen Proving Grounds, and Rocky Mountain Arsenal, to mention a very limited number. The mules in Grand Canyon National Park, under the jurisdiction of the National Park service, are major wastes generators and polluters of Bright Angel Creek. What stimuli might induce waste reduction in such installations as the military complexes in San Antonio? or in the government office buildings in Washington, O.C.? The federal reduction in paperwork legislation seems to have had little effect.

The subsidy provided by the Postal Service for junk mail and catalogs is a major contributor to wastes generation; the subsidy exists to some extent for mailing and completely for disposal. That is, the perpetrator of junk mail pays none of the costs of disposal. A recent study of small town post offices in Vermont indicated that 85%-90% of the substantial amounts of paper wastes generated in those post offices consisted of junk mail tossed away by boxholders. Increasing rates on junk mail to reflect disposal costs, and imposing those costs on senators and representatives --- for their mailings to constituents --- would have a positive effect.

Responses to Stimuli

The responses of some segments of the private sector to one or more of the various stimuli could be characterized as falling into two classes. One represents the application of existing technology, which --- though available and might have been applied and increased net revenues --- had not been applied for various reasons, e.g., inertia, lack of capital, waiting until major overhaul was due, simply insufficient stimulus. For example, according to Kane (op. cit., p. 133), "The steam and condensate distribution system in the mill of the future will not need new technology. All the technology and equipment required is in use today. It will just use all of this technology and equipment in an integrated system." Dioxin generated in pulping is another example. The oxygen supplements, oxygen bleaching, chlorine dioxide substitution for chlorine, oxygen delignification, all were already known before the concern for dioxin suddenly arose. Some mills years before the recent flurry moved in a direction which would and did reduce dioxin problems, even though at the time concern for dioxin was not the stimulus for change. Rather, the shifts related to long-run production costs, product yields, product quality.

The second type of response is to push development of now technology to cope with a new problem. The perception of the problem may not be in relation to reducing generation, but rather in relation to use of residuals now going to the landfill. Development of technology to enable de-inking so as to be able to use laser-Printed computer printout (CPO) waste paper for raw material in papermaking is another example. (See Gilkey, et al., 1988.) Thus, one technological development may drive the development of another technology.

The laser CPO example shows how gross generation can increase at the same time net generation decreases. This is analogous to the trend in water use in several heavy industries over the last three decades,

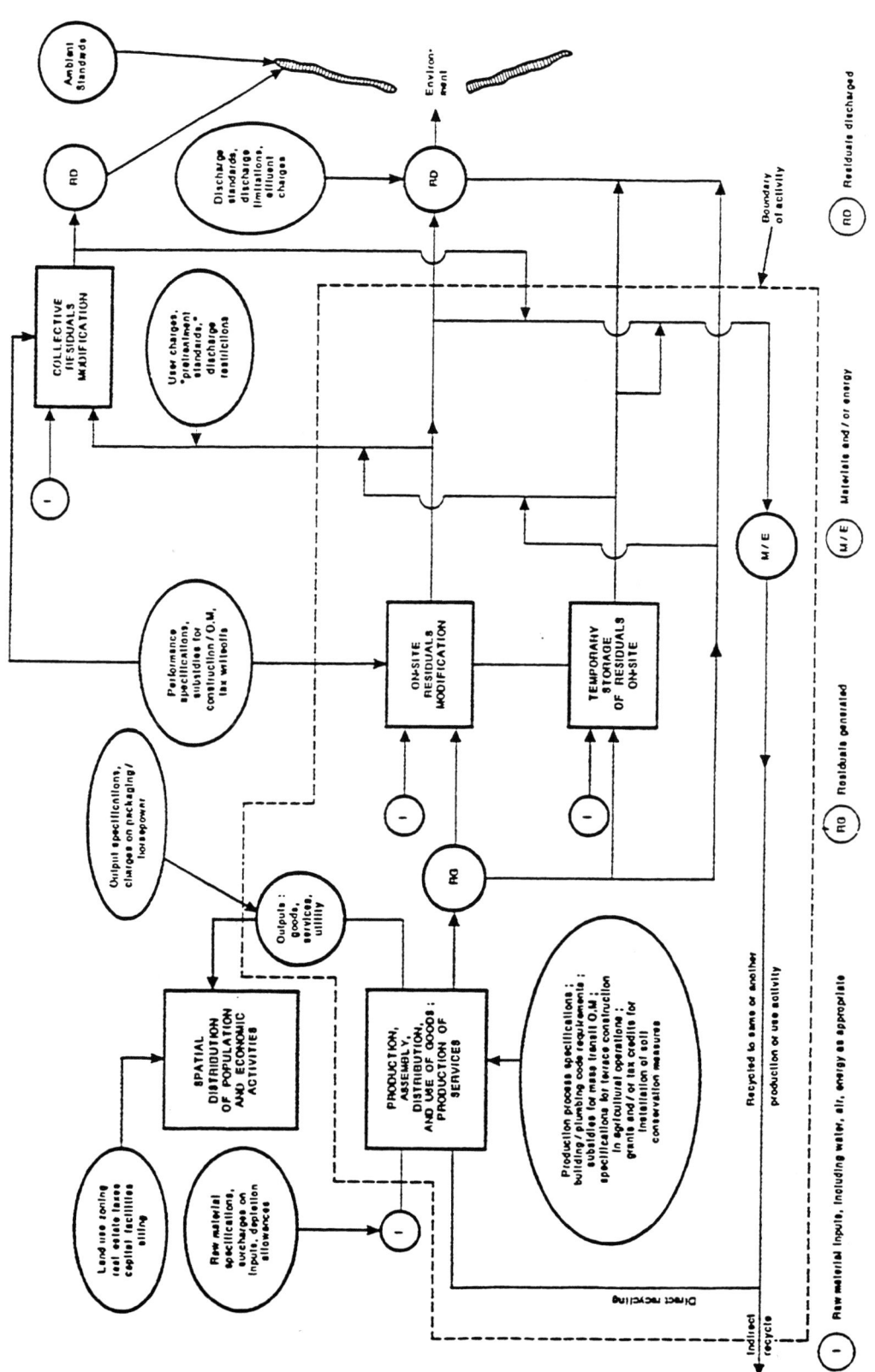

FIGURE 6 Possible Intervention Points for Waste Reduction

pulp and paper for example. Gross water applied per unit of pulp and paper product has <u>increased,</u> as a function of e.g., increased product specifications. At the same time, water intake per ton of product has decreased, because the increase in water recirculated has more than compensated for the increase in gross water applied.

CONCLUDING COMMENTS

Three comments will conclude this rambling discourse.

(1) Waste reduction in a region --- in terms of reducing the net quantity of residuals discharged to the environmental media --- can be achieved by more explicit consideration of possible linkages among intakes and discharges of activities in the region, as suggested by the flow diagram of Figure 7. Developing such linkages among activities within a region is analogous to developing linkages among different units on a plant site, such as a petrochemical plant.

(2) Determining whether or not waste reduction, however defined, could be achieved by any given proposal/technologic option requires analysis of the total system. The basic structure for such a system analysis is shown in Figure 8 for a paper product. Associated wood products are shown, because of the linkages between the two types of outputs. Energy, water, and other inputs at various points in the system would have to be shown, plus transport costs, plus explicit specification of residuals generated at various points in the system- However, one conclusion can be drawn just on the basis of the structure, namely, if product specifications could be lowered, as discussed with respect to consumer paper products, waste reduction in terms of gross generation could be achieved.

(3) The continuing trend for more product proliferation and higher product specifications, taller office buildings requiring ever more sophisticated stool, and more energy intensive activities such as downhill skiing, makes the achievement of waste reduction in American society difficult. That situation will not change until the mores of the society are more in accord with the view attributed to Henry Ford I, "you can have any model and color of Ford car you want as long as it's black and a Model A."

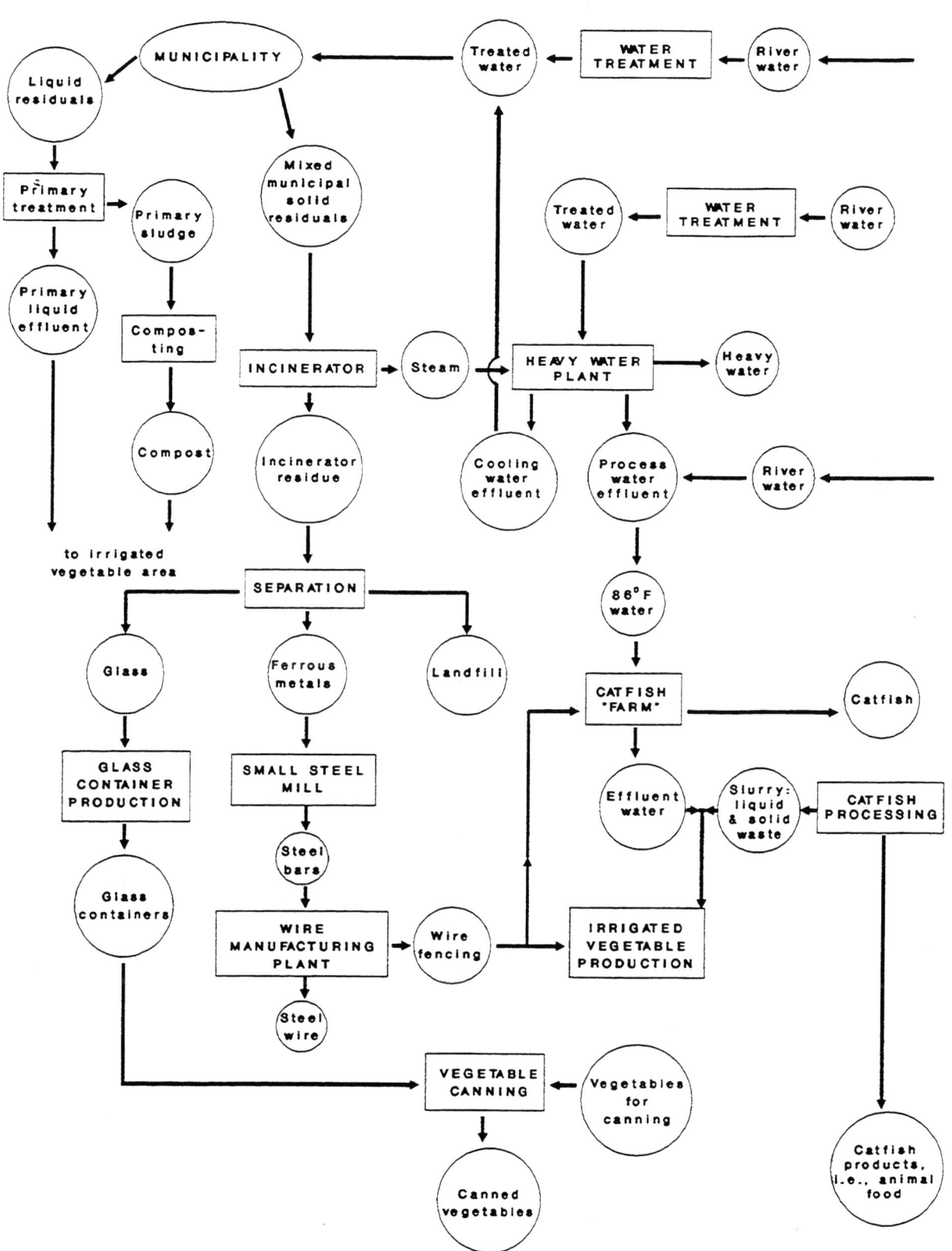

FIGURE 7 Example of Linking Activities in a Region

74

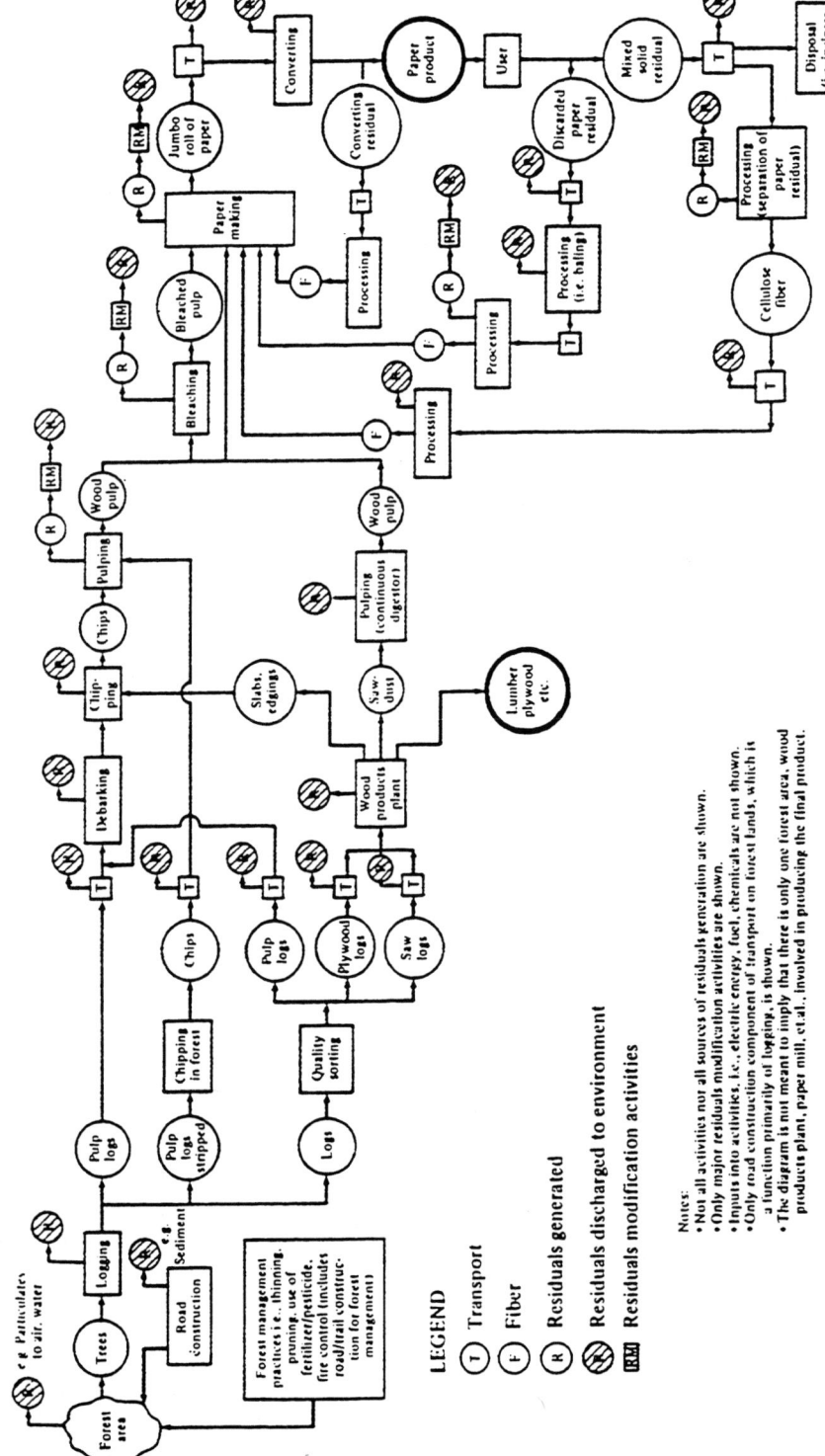

Source: **Kneese and Bower (1979, p. 115)**

FIGURE 8 Components of Systems for Producing a Specified Paper Product

REFERENCES

Anon., 1989a, Outwitting trout with data, New york Times, 19 March, p. 49.

Anon., 1989b, Plant recycles foam products, Waste Age, 32, 2, p.14.

Anon., 1988, Domtar and Dow Chemical to recycle plastic bottles, Pulp and Paper, 62, 11, pp. 34-35.

Anon., 1987, The changing economics of waste management, Chemecology, February, Chemical Manufacturers Association, Washington, D.C.

Anon., 1974, The squeeze on product mix, Business Week, $2312, 5 January, pp. 50-55.

Associated Press, 1989, Bright idea: colored plastic wrap, Post, 7 March, p. C14.

Bower, B.T., Lof, G.O.G., and Hearon, W.M., 1973, Residuals in the manufacture of paper, Proceedings ASCE, 99, EE1, pp. 1-16.

Bringer, R.P., 1988, Pollution prevention plus, Pollution Engineering, 20, 10, pp. 84-89.

Bryan, E.H., 1984, Biobricks become a reality, Water Engineering and Management, 131, 3, pp. 38-39, 59.

Cheremisinoff, P.N., 1989, High hazard pollutants: asbestos, PCBs, dioxins, biomedical wastes, Pollution Engineering, 21, 2, 58-65.

Day, J.W., 1974, Can we afford to continue cosmetizing paper products during a time of shortages?, Chem 26, February, pp. 21-22.

Dumbrowski, C., 1989, Landfill ban prompts now hazwaste treatments, World Wastes, 32, 2, pp. 24-25

Economic Research Service, 1988, Food costs .... from farm to retail, U.S. Department of Agriculture, Washington, D.C.p pp. 8.

Gilkey, M., Shinahara, H., and Yoshida, H., 1988, cold dispersal unit boosts de-inking efficiency at Japanese tissue mills, pulp and Paper, 62, 11, pp. 199-103.

Hahn, G.E., 1989, Effective condensate management can cut energy, maintenance costs, Pulp and Paper, 62, 11, PP130-133.

Hiatt, F., 1989, The motherland of invention, New York Times, 30 March, pp. A33, A35.

Hudson, J.F., Lake, E.E., and Grossman, O.S., 1981, Pollution-Pricing, Lexington Books, O.C. Heath and co., Lexington, Massachusetts.

Johnson, B., 1989, Kirkland Recycling program gets unexpected success, World-Wastes, 32, 2, pp. 26-29.

Kane, J., 1989, Information control assists in production, customer processing, Pulp and Pacer, 63, 2, pp. 94-95.

Kneese, A.V., and Bower, B.T., 1968, Managing Water Quality, Johns Hopkins University Press, Baltimore, Maryland.

Kneese, A.V., and Bower, B.T., 1979, Environmental Quality and Residuals Management, Johns Hopkins University Press, Baltimore, Maryland.

Lowe, K.E., 1973, Too much quality in paper and board?, Pulp and Paper, 47, 12, p. 62.

Rabine, J., and Winton, J.M., 1974, Energy and the product mix, Chemical Week, 115, 22, pp. 29-32.

Tucker, R., and Sutherland, G., 1988, Guidelines allow the evaluation of capital project design effectiveness, Pulp and Paper, 62, 9, pp. 91-95.

ACCOUNTING AND THE ENVIRONMENT:
Patching the Information Fabric

Rebecca Todd

Assistant Professor of Accounting
New York University

I want to thank you for inviting me to this conference. It is an honor for me and I'm very grateful.

We have set ourselves a considerable task: to propose important avenues for research over the next few years which will enable us to strengthen the delicate fabric supporting the environment. It is my conviction that a number of separate economic changes which have occurred recently and are, in fact, underway at this moment will make our work simpler and clearer and vastly improve the likelihood of success. Indeed, the timing could scarcely be better.

At the risk of stating what is obvious in any case, all persons have a stake in the health of the environment and the conservation of scarce resources. However, my proposals assume that of all possible constituencies, two, managers of firms, and capital market participants, have unique positions, which enable them both to control how efficiently productive resources are used, and to improve and preserve the quality of the environment. Of the two, I consider corporate managers to be the ultimate fiduciaries of the environment. The topics for research which I propose are aimed at finding ways to remove institutional barriers which currently hinder the functioning of both of these groups.

Specifically, I will propose that we need to do two things: (1) close some gaping holes in the traditional accounting information fabric which allow critical information to go uncaptured; and (2) support users of this information with a regulatory framework which provides a level playing field for economic decision-makers. Simply put, people can't act if they don't have adequate information, and they won't act voluntarily if they will bring harm to themselves by doing so.

The proposals which will be elaborated on below will be to develop means to:

(1)   revise our traditional production accounting methods which capture and apply to products those costs which occur in production, but which cease when the finished product leaves the shop floor, disassociating spillovers and contingent costs from the products which produced them;
(2)   provide financial and non-financial environmental information to capital market participants, i.e., equity investors and creditors, whose business it is to evaluate the relative profitability of firms and the associated risk in the setting of costs of capital supplied to firms.

As will become clear later, these are not independent but are different aspects of the same fundamental issue. Moreover, the ideas are not new. What is new is that the legal, regulatory, and accounting domains have undergone significant changes in recent years, particularly in the past two, which increase the probability of achieving these objectives, even making them imperative. These will be discussed in context below.

The remaining sections of this paper will provide relevant background on the nature of accounting and the regulatory bodies which control accounting output, as well as discussion of each of the proposals with recent examples.

Nature of Accounting

Accounting, to use the philosophy of science term, is a "reconstruction" of an enterprise, a financial model. It is the reflection of a firm from a financial mirror. In general, then, only economic events which can ultimately be defined in financial or dollar ($) terms will be captured by the system. However, the system is capable of capturing any information which the management of a firm deems useful, and indeed, the extension of the accounting system to its larger family, the management information system, does exactly that.

It is considered to be a desirable trait of accounting information--the output of the accounting system--that the information be neutral, or unbiased in order to provide the best basis for economic decision-making. That is, accountants should be detached, impartial observers of the operations of the firm and not proactive instigators or advocates of various courses of action. Such activity is thought-to be the appropriate domain of the managers of a firm. I trouble you with this observation to make clear that the information which is captured by the accounting system is determined largely by persons and agencies outside of the accounting department.

Managers can direct that any desired information be collected, analyzed, aggregated, summarized, and reported to them, subject only to the usual benefit/cost and other feasibility constraints. For example, management might desire that the production schedules and total sales for the next two years for each of their competitors be reported to them, but this would run into a feasibility constraint!

Information provided to outsiders includes management's voluntary disclosure which common wisdom suggests is that which is favorable to the firm, although some interesting and puzzling counter-examples exist. However, the vast proportion of information which is available to investors, creditors, and others is mandated by two regulatory bodies, the Financial Accounting Standards Board (FASB), the private rule-making organization which provides guidelines for the public accounting profession, and the Securities Exchange Commission (SEC), which has the primary responsibility for overseeing the equity and debt securities markets and seeing that "adequate" information is provided to market participants.

Thus, if we are to change the ways in which accounting information is collected and reported, we must convince one or more of these three groups that it is desirable to do so.
one brief word, an opinion, regarding benefit/cost analyses. Such analyses are the usual analytical and economic-information-generating system for government agencies as well as corporations. Much very important work has been done here, but as a tool for persuading rational corporate managers of the rightness of a course of action, they suffer from a taint well-understood by managers.

Benefit/cost analyses are themselves costly commodities and are rarely ordered up in a vacuum. One usually has some objective or outcome in mind before the resources are committed and expended on the analysis. Thus, suspicion will be immediately aroused in the beholder of the information that the assumptions, analyses, and conclusions may well be scented with the perfume of the desired outcome! This will not come as news to anyone here. It is my view that the considerable efforts which have been expended on these analyses have not "gone to waste" but will become an invaluable resource themselves when other incentives are in place. I believe that they rise to their highest and best use not as motivators, but as decision support aids. That is, how shall we best choose among alternatives given that we've already committed to the battle.

Indeed, the world of private accounting is peopled with thousands of experienced, superbly-trained and highly analytical cost accountants. What is needed is to persuade management to unleash them on the problem.

Proposal I--Revisions to Product Costing Systems:

Full-Costing for the Twenty-First Century

Before we launch directly into the issues here, let me provide a short primer on accounting. Accountants have a number of tools at their disposal, or images, which they may present to managers or outsiders of the firm. Of these, two are of central importance: the Balance Sheet, and the Income Statement. The balance sheet is an assay of the resources which a firm commands, the assets, and who has claim on those assets, the creditors and owners (or shareholders). The balance sheet provides a snapshot at a specific time of the financial health of an enterprise. The Income Statement is the linkage between balance sheets, and monitors the inflows and outflows of resources. In short, it takes the pulse of the firm. As products are being manufactured, accountants collect the costs which are being added to the item, aggregate them, and then match these costs to the sales price of the item when it is finally sold. The difference between sales and costs is what accountants term Income, a somewhat different meaning from the popular usage of the term.

In the old days, before we became aware of how fragile the environment truly is, of how dependent upon it we are, and of how scarce our resources are, cost accountants, at the behest of management, developed systems of tracking and aggregating costs to facilitate optimal investment and production systems. As dumping of wastes and other leftovers was a costfree activity, there was no need to capture these costs. Indeed, as I have noted above, inventory costing stopped when the product left the shop floor.

Other costs, such as marketing expenses have traditionally been accounted for separately and not charged to product. As waste control and disposal have become increasingly expensive and must be recognized, they have generally been treated as period costs, that is, as general expenses. Most usually, this cost will be accounted for as general overhead, either being allocated between cost of goods sold and ending inventory at the end of a reporting period, or dropped entirely into cost of goods sold.

Let us consider some simple cases.

If we should not be required to account for the costs of waste and pollution, that is, if these spillovers are free to the firm, then the costs to the firm are understated, the income is overstated, and society at large, which must ultimately bear the cost of the cleanup has effectively paid a subsidy to the firm to foul the environment and be profligate with resources.

A somewhat notorious case of the failure to adequately internalize the costs of products, although not an environmental problem in the usual sense, is that of the Savings and Loan industry. The problem here was "waste" of capital, and the "cleanup" is now estimated to cost each taxpayer in the United States about $500! It is an apt example because the source of the problem is the same one which we have been discussing, a rip in the cost accounting fabric through which billions of dollars of scarce capital poured. It is also particularly appropriate because it represents a regulatory failure as well. All of the expected safety nets collapsed and losses rivalling the national debt have fallen on the shoulders of everyone outside of the Savings and Loan industry. We will return to this example later.

Another aspect of cost capture which we should consider at this point is the accounting Contingency. Under Statement of Financial Accounting Standard (FAS) #5, "Accounting for Contingencies"[2], if a firm has already suffered a probable loss or impairment of existing assets, then some form of recognition of this expected loss must be made in the income statement and/or balance sheet. In general, this rule has been applied in such cases as large lawsuits (e.g., Pennzoils $10 billion dollar judgement against Texaco) large catastrophic losses in excess of expected insurance coverage (e.g., the mid-air collision over Los Angeles some years ago), and large expected losses on contracts which have not yet been completed (e.g., commodities forward contracts for food processors).

The contingency rule has not been applied to the usual run-of-the-mill spillover. The argument has been made that the requirements are not met, i.e., we haven't yet suffered impairment of assets or incurred a loss, and won't until lawsuits are filed or regulatory action is taken.

Change is in the wind. A recent flurry of court judgements, particularly in New Jersey, have ruled that insurers cannot be held liable for losses and damages arising from ordinary business decisions which have been in place over a period of years. This somewhat simplified summation of a series of cases has suggested to the financial community in New York that firms may no longer be able to rely on the insurance company to cover the breach. The argument goes that such losses were reasonably foreseeable and, as a consequence, usual business expenses. Thus, firms will stand alone in the liability for losses and damages, should they occur.

Another gap appears in the accounting fabric. The contingency rule has not been extended to cover such cases before the damage case is filed, insurance companies may not be liable, and firms are not currently permitted under the accounting rules to set aside General Reserves for Losses. In short, there is no cover for the firm.

We would seem to have two choices: (1) to seek ways to begin capturing this "uncovered" cost in the regular product costing system and financial disclosure systems; and/or (2) immediately begin to reduce the waste and environmental hazards at the source, the ultimate solution. It is a curious economic fact that the former will almost certainly produce the latter!

I really must stop to note at this juncture that a heartening number of firms have already begun to take these steps. Approaches have included inventorying products, associated wastes and toxic substance "leftovers," and investigating the feasibility of conversion of these leftovers either to salable products, recycled factors, or neutralized substances. The managers have done so unilaterally, without the sort of incentives that most of us would like to see in place, save a very long vision into the future.

A case which will be familiar to most of the conference participants is what happened to the textile and steel industries in the United States when foreign competition became a serious factor. The argument went at the time that retooling to develop more technologically advanced and more efficient, and thus less costly, production was too expensive for the firms to bear. Moreover, strong arguments were made that foreign governments were subsidizing the competition when such firms already had an advantage in the form of lower labor costs.

Thus, a system of investment tax credits was put in place coupled with a complex web of tariffs, quotas, and other restrictions, to shield the industries from the "predatory" competition and to grant the firms time to retool. You will recall what happened. Nothing! At long last, Congress grew weary of subsidizing the industries, the country entered a deep recession, and the trade restraints were lifted. The firms were left to swim alone in the markets. Dire predictions were made and the death knell was sounded. Indeed, it was not a pretty sight. Some firms did go under or were absorbed in extremis by other organizations. However, death notices were, in general, premature. Within a few short years, the survivors had retooled, pared the fat from their operations, and repositioned their products to supply what the markets were demanding, and the industries have begun a new era. Market pressures and incentives accomplished what no amount of sheltered encouragement could achieve.

In the long run, say some decades or more, I have every expectation that technology will solve many of the environmental problems. But what short-term steps can we take to capture these costs and provide the necessary incentive to spur the changes? Possibilities include but are not restricted to:

(1) qualitative disclosure of waste materials and other hazardous wastes produced in manufacturing processes as a minimum disclosure;
(2) a contingency rule type of disclosure where aggregate minimum estimates are made across the spectrum of products produced by a firm;
(3) development of definitive costing standards to estimate and capture specific costs of potential environmental losses.

In the last six months or so, the major international accounting conferences have focussed on a single broad issue: the inadequacy of current cost accounting standards for product costing in the modern manufacturing environment. The impetus for this reexamination is the rapid modernization and automatization of manufacturing processes under the extreme pressure of foreign competition. The buzzwords of the new technologies are familiar, just-in-time inventory control, robotics, etc. A central feature of the new technologies is that labor is becoming an increasingly smaller factor in production, in many cases amounting to less than 20% of total product costs. Our antiquated cost accounting systems are still using primarily labor-based cost standards, allocating overhead and other fixed charges on the basis of labor hours or labor costs, etc. At the European Accounting Association meetings a few weeks ago in Stuttgart, Germany, the central theme of the conference was to modernize the costing systems to better reflect the new realities of the shop floor.

A related but more subtle result of the new technologies is that the newer systems will be less flexible and thus less susceptible to modifications once the system has been installed. The important point for us is that fewer opportunities will be available once plant construction is completed and production has begun to reduce the environmental wastes produced. Thus, we may increasingly find that our last best hope of modification occurs at the plant feasibility and planning stage. This topic I will defer to the engineers and technology experts among us.

The final point I wish to raise in this section is that time is of the essence in seeking to bring environmental costs under the firm's costing umbrella. The courts have spoken, technology is rapidly changing, and traditional cost accounting is now loose from its moorings, is drifting in a state of flux and is most susceptible to conversion to a full costing which includes the environment.

Proposal II: Capital Markets and the Valuation of Firms
--Levelling the Playing Field

What will have become abundantly clear to any corporate manager by this point is that restructuring of cost accounting standards alone is inadequate. That is to say, developing new techniques for managers to use internally to the firm will not achieve the desired goal of reducing environmental waste. The reason is that a manager has no incentive to impound these costs for external financial disclosure, reduce the firm's income, and unilaterally suffer the capital market's displeasure at his poor operating performance, relative to that of his competition. To move from the internal managerial costing tool to an external disclosure requires that the playing field be levelled. That is, that the rules be applied uniformly to all publicly-traded firms and, by regulatory extension, to all private firms as well. Football is not played on a mountainside.

First, a bit of background. The capital markets, both equity and debt, have enormous power. This power derives from their function, the setting of the cost of capital and the terms of capital acquisition. Capital, the funds necessary for fuelling ongoing operations as well as future growth, is provided to firms at a cost. This cost is termed the "return" to the capital provider. In general, the cost is a function of the market's expectations about the future profitability of the firm and the riskiness of the resulting future returns to the capital providers.

Capital providers are, in general, risk averse. Thus, if two firms are compared and one appears to represent greater risk than the other, the suppliers of capital will demand a higher return for bearing the risk. This translates to a lower price for the equity shares trading in the market, and a higher interest rate for debt capital.

Now, if the costs of environmental waste are impounded in the earnings of a given firm, reducing the earnings, the market's expectations of future returns will be reduced, at least for the near horizon, and the price of the firm's shares will be adjusted downward. Unless the firm chooses to cut its dividend, an event which would likely be regarded negatively, its effective cost of capital will rise.

Let us consider a second case. If a firm discloses, or it becomes otherwise known in the markets, that the firm has a potential loss exposure resulting from environmental waste, the market is aware that insurance may not cover the loss, and that the firm will ultimately stand liable for the loss, the market will conclude that the risk of the firm has risen. As shareholders are risk averse, they will value this riskier firm at a lower amount, and again, the cost of capital for the firm rises.

Finally, let us assume that very limited information is available about the potential losses a firm may suffer. As anyone who was alive on October 19, 1987 knows, the market is a skittish animal. Great uncertainty is the worst possible case for the market. Firms for which the future is thought to be very uncertain are considered speculative in nature and the market is ruthless in its assessment. Such firms may effectively be barred from the capital markets by prohibitive capital costs.

If a firm unilaterally decides to disclose the environmental risks in its product portfolio, the market will likely look unfavorably on the firm. An argument can be made that if one firm in the industry has chosen to "come clean," the firms which have not will be subject to greater uncertainty and will experience even greater approbrium. Such a result may indeed occur, but the cost for the first firm will be high in any event.

The case of commercial banks makes this abundantly clear. Over the last twenty years, the banks have invested in portfolios of increasingly risky assets including large amounts of lessdeveloped-country debt (LDC), off-the-books guarantees for substandard domestic municipal bonds and other debt, commercial mortgages in energy-producing states, and more recently, leveraged-buy-out debt (LBO). While the market has been aware that this has been occurring, the financial disclosure of the extent of these holdings and the losses incurred in the portfolios has declined, in some cases precipitously. The market's judgement has been devastating. The average bank's cost of capital is approximately twice that of other Standard & Poors firms, and for some large money-center banks the price/earnings ratio is approximately three. Implied costs of capital range upwards of 20-30%. The distressing part is that the markets have not chosen to distinguish cleanly between the excellent banks and the worst offenders, a sort of "guilt-by-association."

This is directly analogous to the Savings and Loan industry which was cited earlier. In an accounting disclosure vacuum, all suffer.

Several points should be made.

First, if disclosure is to be made, it must be made uniformly by all firms under carefully considered regulatory mandate. This will assure that no firms are subjected unilaterally to the market's wrath.

Second, the disclosure should be as clear and unambiguous as it can be made in order to dispel unnecessary uncertainty in the minds of shareholders.

Finally, we may well have passed the point where such disclosure is an alternative option. The financial analysts' rumblings about the recent court cases are but one example of the concerns extant. If the market suspects that firms have undisclosed risks, and that the risks are possibly quite high, the market will assume a worst case situation which is undesirable for all firms.

Objections can be raised to the proposed financial disclosure. This will not be the first time, but I believe the arguments will diminish in the face of the alternatives. First, protests will be made that some firms will be badly and unfairly hurt. As with the steel/textile example, some firms will indeed suffer. The good news is that those firms which have begun early to get their environmental houses in order will be in a first rate position relative to the competition by virtue of their careful and foresighted stewardship.

Second, some industries, regardless of their efforts to avoid and prevent waste production, are inherently "dirtier" than others. A chemical company is not a David's Cookies whose major hazardous waste is a bad pecan or two! True. Again, however, the market will eventually make its judgement and we would prefer it to be a carefully considered and reasoned decision based on clear and unambiguous information.

Additionally, measurement is a nontrivial problem. It always is in accounting. We feel comfortable with what has become familiar and the usual operating terrain. The fact is, however, that much of what we measure and report is subject to vast uncertainty and measurement error. Depreciation of fixed assets is familiar, is usually a large number, is highly discretionary, and carries enormous uncertainty in rapidly changing operating conditions.

Another example, an especially complex one, may serve to illustrate. For years, firms have provided or promised pension benefits to employees. The prospective costs and liabilities for these benefits were enormous. Numerous abuses resulted in unfunded pensions and the finding that only a tiny proportion of employees ever received the promised pensions. The final straw was the pressure on the Social Security system brought about by the private sector pension system failure.

A network of Congressional laws, the Employees Retirement and Income Security Act (ERISA)[3] of the early 1970's, tax incentives for those firms which met the strict regulations for qualified pensions, and a series of financial disclosure rules (the most recent is the FASB's Financial Accounting Standard #87 [4]) has effectively closed the major gaps in pension accounting and funding.

I consider this case particularly appropriate because the attributes of this instance map almost completely to the difficulties we see in environmental accounting and disclosure.

First, and most prominently, are the uncertainties about the eventual costs. How are we to decide how much a young worker who joined the firm a year ago will eventually receive in a pension? Will he be with the firm? Will he die in the interim? Will he be promoted? What salary increases will occur in the interim? How will changes in pension plans be made over the next several decades. The accounting rules have not provided perfect solutions to these problems, but with the aid of actuaries, demographers, labor economists, etc., gradual iterations of the rules have produced some workable compromises. Second, the costs are indeed enormous and firms had to choose whether to suffer expert labor losses if they decided not to provide such plans, or to endure the market's discipline if they disclosed the large costs. There was really little choice.

Third, many firms were in a decidedly inferior position relative to other firms. Again, those who had begun this stewardship early were the clear winners.

This solution is not yet fifteen years old and the problem has largely abated. The funds transferred to trusteed accounts have grown rapidly with the bull markets. Most firms now have their cost problems under control. Perhaps most importantly, the issue is no longer at the forefront of concern as it was a decade ago.

The discipline of the markets will work. It will work quickly, and it will work surely. The question is whether we will be prepared to deal with it in an orderly and well-considered fashion.

I propose that research be directed at developing an integrated network of provisions along the lines of the ERISA model. I recommend that initially we take the viewpoint that all waste items are candidates for consideration. Specifically, we should examine:

(1)   which disclosure components are essential;
(2)   which methods of capturing the costs and reporting the findings will best reflect the underlying economic exposure of the firms;
(3)   the feasibility of structuring and implementing tax incentives in conjunction with disclosure compliance;
(4)   the possibility of funding of a portion of the exposure.

The incentives here are clear. We may soon find that we can focus all of our efforts on improved technology. Such has happened in the case of the implementation of rules requiring current financial disclosure of foreign currency transaction losses, FAS #52[5]. The losses have rapidly become a relic of the past as firms have learned to structure their contracts to avoid producing the losses in the first place and to hedge what remains.

Public financial disclosure is not only a powerful tool. It can make all others obsolete!

ENDNOTES

1.   Excellent research, discussion, and review of the problems leading to the savings and loan association crisis are to found in An Analysis of the Causes of Savings and Loan Association Failures, by George J. Benston, Salomon Brothers Center for the Study of Financial Institutions, Graduate School of Business Administration, New York University (1985), and Thrift Financial Performance and Capital Adequacy, Federal Home Loan Bank of San Francisco, Proceedings of the Twelfth Annual Conference (1986).
2.   Financial Accounting Standards Board (FASB)--"Accounting for Contingencies," Financial Accounting Standards (FAS) #5, (Stamford,CT:1975).
3.   The Employment Retirement Income Security Act (1974).
4.   FASB--"Employers' Accounting for Pensions," FAS #87 (1985).
5.   FASB--"Foreign Currency Translation," FAS #52 (1981).

ACCOUNTING AND THE ENVIRONMENT:
Patching the Information Fabric

General Objectives

A. Close gaps in the traditional accounting information fabric which allow critical environmental information to go uncaptured; and
B. Support users of this information with a regulatory framework which provides a level playing field for economic decision-makers.

Basic Assumptions

A. Potential fiduciaries of the environment can't act if they don't have adequate information; and
B. They won't act voluntarily if they will bring harm to themselves by doing so.

III. Research Proposals

Proposal I--Seek methods to revise our traditional production accounting methods which capture and apply to products those costs which occur in production, but which cease when the finished product leaves the shop floor, disassociating spillovers and contingent costs from the products which produced them.

Possibilities include but are not restricted to:

(1) qualitative disclosure of waste materials and other hazardous wastes produced in manufacturing processes as a minimum disclosure;
(2) a contingency rule type of disclosure where aggregate minimum estimates are made across the spectrum of products produced by a firm;
(3) development of definitive costing standards to estimate and capture specific costs of potential environmental losses.

Proposal II- Develop prototypes of financial and non-financial environmental information disclosure to capital market participants, i.e., equity investors and creditors,, whose business it is to evaluate the relative profitability of firms and the associated risk in the setting of costs of capital supplied to firms.

Possible steps include determining:

(1) which disclosure components are essential;
(2) which methods of capturing the costs and reporting the findings will best reflect the underlying economic exposure of the firms;
(3) the feasibility of structuring and implementing tax incentives in conjunction with disclosure compliance;
(4) the possibility of funding of a portion of the exposure.

SOURCE REDUCTION: WHAT IS IT AND HOW CAN WE ACCOMPLISH IT?

Katy Wolf

INTRODUCTION

The hazardous waste community has spent much time in the last several years debating the definition of source reduction and its role in the waste management hierarchy. New phrases have been coined to better represent our enlightened understanding. Indeed, source reduction is no longer the accepted phrase; rather, we now speak of pollution prevention to symbolize the insight that source reduction applies not only to waste, but is multimedia in nature. There appears to be a consensus now that pollution prevention is the correct term but there is still not a consensus on what this term embodies. Should source reduction--or, pollution prevention as it is now called--be defined to include on-site treatment? Should it be defined to include on- or off-site recycling? There is not yet agreement on this question and debate on it will continue.

A related issue that also remains unresolved is how to measure progress in source reduction. A variety of schemes have been proposed but none of them appears to be generically applicable. In some instances, there will be agreement that source reduction has been achieved and that it can be measured. In general, however, we may have to accept that there is no universal approach and agree to subjectively judge the level of source reduction on a case-by-case basis.

Many of us are anxious to go on the political record in support of source reduction. Virtually no one is against the idea of source reduction, so there is no risk in calling for it to be practiced. One bill that would require industry to report on source reduction activity has been proposed. It has widespread support from many, including several industrial firms. A second bill that would require source reduction to be implemented has also been proposed. It applies a uniform standard across many industries and would give firms a decade to accomplish the mandated source reduction.

This paper focuses on four issues facing the source reduction community today. First, it describes the continuing debate on what constitutes source reduction. Second, drawing on a case study of four plants producing or using a substance classified as hazardous, it discusses the difficulties in measuring source reduction. Third, again using the case study for exposition, it describes and analyzes the impact of two proposed bills that would regulate source reduction. Fourth, it suggests a nonregulatory approach to improved management of hazardous substances.

The findings of the analysis are that there is no generic method for measuring source reduction. The complexities of industrial processes and the disparate effects of different substances on human health or the environment suggest that measurement can be done only on a case-by-case basis. Even then, there may not be agreement on whether or how much source reduction has been accomplished. The second conclusion is that a regulation requiring information collection or a regulation requiring source reduction would not make sense. We do not know enough about how to measure source reduction to analyze the data or to judge whether source reduction across or among industries has been accomplished. Furthermore, such regulations would impose a disproportionate burden on small and medium sized firms and some of them could go out of business because they would not be able to comply. A more productive way to move toward improved hazardous substances management would be to establish technical assistance groups to help small and medium sized firms in a plant-by-plant effort. Such groups would need good knowledge of industrial processes and an understanding of the tradeoffs involved in adopting alternative chemicals, products and processes.

Better protection of human health and the environment should be the aim of us all. Source reduction, a laudable goal, may help to achieve it. Its consequences, however, are more complex than the source reduction community as a whole is willing to acknowledge today. Many in this community believe that knowing a few concepts about the waste management hierarchy qualifies them to assist industry in source reduction. In fact, it is important to take into account the tradeoffs in moving from one hazardous substances management regime to another. At best, if we regulate source reduction before we know what it is or how to measure it, we will add to the bureaucracy of existing inefficient and conflicting regulations. At worst, if we do not take into account the tradeoffs we may actually injure people or damage the environment.

THE DEFINITION OF SOURCE REDUCTION

The idea of the waste management hierarchy grew out of the unavoidable conclusion that unrestrained land disposal had been a mistake. This hierarchy includes the various elements of waste management and is commonly defined as follows: source reduction recycling physical, chemical or biological treatment thermal treatment land disposal. This hierarchy is an ordering of hazardous substances management options according to their presumed environmental consequences with those at the top posing less of a threat than those at the bottom. The concept of the hierarchy is accepted by most people today. Source reduction--the top element--is preferred over all the elements lying lower in the hierarchy because it is assumed to be environmentally safer. Source reduction includes activities that reduce the use of hazardous substances at the source before wastes are generated. In a particular plant, such measures should be identified and implemented to the maximum extent possible. Once source reduction opportunities have been exhausted, the waste manager moves to the next element, recycling. In-process recycling is preferred over on-site recycling which is in turn preferred over off-site recycling. Once all recycling opportunities have been exercised, the manager may move to various types of treatment. Again, on-site treatment is preferred over off-site treatment. Once treatment has been practiced to the extent possible, the manager may incinerate the waste. Only after all these options have been adopted should land disposal of the waste be considered.

There are a number of players in the source reduction game and the term has a range of definitions. On the one extreme, The Congressional office of Technology Assessment defines a term called waste reduction as "in-plant practices that reduce, avoid or eliminate the generation of waste so as to reduce risks to health and the environment" (OTA, 1986). This definition does not include on- and off-site recycling as legitimate source reduction activities. In-process recycling, or closed loop recycling that is part of an industrial process, is the only form of recycling included in the definition. On the other extreme is a term accepted by the Environmental Protection Agency (EPA) in the past and by many in industry called waste minimization. It generally includes source reduction,' on- and off-site recycling and treatment. The EPA recently established a new Office of Pollution Prevention which was to take a primary role in source reduction. At first, the office seemed inclined to adopt the OTA definition. The final official definition, however, was broader; it included on- and off-site recycling but excluded treatment of any kind (Fed Reg, 1989). Various other players define source reduction according to their viewpoint.

In an earlier paper, I described the controversy that surrounds the definition of source reduction and its role in the hierarchy (Wolf, 1988). There appears to be a consensus that source reduction should occupy the preeminent position at the top of the hierarchy. There is considerable disagreement, however, on just exactly what adopting it "to the maximum extent possible" really means. As I discuss in the earlier paper, one extreme view is that it means that source reduction should be adopted until it is no longer technically feasible. Another extreme view is that source reduction should be adopted until it is no longer cost-effective compared to end of pipe treatment technologies already in place.

There is and will be a continuing debate on the hierarchy and on what constitutes source reduction, waste reduction, waste minimization or pollution prevention. There may never be agreement on the best definition and it probably does not matter. As I show below, whatever term is used, there is presently no good method for measuring it or indeed, for deciding what it is. Throughout, I use the term source reduction to encompass actions that lead to better protection of human health and the environment.

THE CASE STUDY

To explore the questions that arise in measuring source reduction, let us consider four plants described in Table 1 that produce or use methylene chloride (METH), a chlorinated solvent. In certain laboratory tests, this chemical has caused tumors in animals. Although the International Agency for Research on Cancer (IARC) classifies METH as a possible carcinogen, EPA defines it as a probable human carcinogen. EPA and California are considering regulating it as a toxic air contaminant, the Consumer Product Safety Commission has a labelling requirement when it is used in certain products, it is regulated under Proposition 65 in California, and spent METH is considered a hazardous waste under RCRA.

In the first plant, a chemical production plant, METH is produced. In the second plant, METH is used in a metal cleaning application. In the third plant, METH is used as an ingredient in an aerosol paint. -In the fourth plant, METH is used as an auxiliary blowing agent in the production of flexible slabstock foam. Table 1 provides a rough materials balance for the four plants. It lists the quantity of METH brought onto the site, released to the various media, and sent off-site in the product.

These plants have very different operations as indicated by the values of Table 1. In Plant #1, the METH producer brings in raw materials and produces METH. In this process, only small releases occur. One thousand pounds of METH is released to the air because of fugitive emissions from valves, pipes and storage tanks. An even smaller amount of METH--500 pounds--remains in the still bottoms from the final distillation at the back end of the reactor. An additional 250 pounds is released to the water. The producer runs a reasonably efficient operation; out of 25,000 pounds of production, the losses are only 7 percent.

The second plant, plant #2, makes lighting fixtures from various metals. These metals arrive at the plant with an oil on them to prevent corrosion in transit and they are degreased prior to machining. After machining, the parts are cleaned with METH again to remove metal fragments, greases and oils. Degreasing is a dispersive process. METH is volatile and most of it--18,875 pounds annually--is emitted to the atmosphere. Occasionally the degreaser must be cleaned out and the sludge, containing 6,250 pounds per year of METH, is sent off-site to a recycler. None of the original 25,000 pounds of METH brought on site goes out on the product; all of it is lost in the degreasing process.

Plant #3 purchased 25,000 pounds of METH for use in an aerosol paint formulation. A small amount of METH--500 pounds--is released to the air during the aerosol can filling operation. An even smaller amount of METH, 250 pounds, is lost to the water when the cans are tested for integrity after filling. The vast majority of the METH, 97 percent of that brought on site, goes out-in the paint product.

Plant #4, the foamer, uses METH as an auxiliary blowing agent in the production of flexible slabstock foam used in furniture, bedding and carpet underlay applications. The function of the METH, in this case, is to expand the foam cells so that the foam has buoyancy. Virtually all of the METH is emitted to the atmosphere within a few days of the foam production process; none of it goes out in the product.

REPORTING RELEASES

Under Title III, Section 313 of SARA, businesses falling in certain manufacturing SIC codes must report to EPA if they produce 25,000 pounds annually after 1988, or use 10,000 pounds annually of a listed substance (Fed Reg, 1988). The list of more than 300 substances includes METH and the four plants in Table 1 are subject to reporting. Under the statute, however, all of the data in Table 1 would not be reported. In fact, only the middle three pieces of data--the air, water and hazardous waste releases--are covered by Section 313.

Let us compare in more detail the plants of Table 1 and see where it leads. Plant #1, the producer, is very efficient. In chemical production, the yield is maximized by minimizing losses and, indeed, in this plant, releases are small. Once the METH has been produced in plant #1, it is sold into various markets like plants #2 through #4. The light fixture manufacturer has large losses of METH. In fact, all of the METH that is purchased from the producer is ultimately lost to the atmosphere. In this particular application, the METH is used dispersively. The same holds true for the foam producer in plant #4. This

Table 1

MATERIALS BALANCE FOR METH PLANTS

QUANTITY OF METH (POUNDS/YEAR)	PRODUCER PLANT #1	LIGHT FIXTURE MANUFACTURER PLANT #2	PAINT FORMULATOR PLANT #3	FOAM PRODUCER PLANT #4
Brought on-site	0	25,000	25,000	25,000
Released to air	1,000	18,750	500	25,000
Sent off-site as hazardous waste	500	6,250	0	0
Released to water	250	0	250	0
Sent off-site as product	25,000	0	24,250	0

SOURCE: Author's estimates.

NOTE: The values are only representative of actual values and are used for exposition purposes. In fact, METH production plants likely have much lower losses than indicated in the figures for Plant #1.

plant purchases the METH from the METH producer and all of it is lost to the air. Plant #3 has much lower losses. As was true for the producer, the paint formulator wants to maximize the METH that goes out in the product. The losses to the atmosphere are accordingly minimal.

In examining the figures of Table 1, we could conclude that plants #1 and #3 are efficient in their use of hazardous substances and plants #2 and #4 are not. Of course, this conclusion is silly because the different operations have different characteristic. Indeed, all of the METE the producer in plant #1 makes is ultimately released to the atmosphere, although not from his plant. The producer sells the METH to the other three plants and it is released according to the characteristics of release of plants #2 through #4. Plants #2 and #4 release it directly during their manufacturing operations. Plant #3 puts it in a product that consumers and commercial painters buy. When they in turn use the product at thousands of locations across the country, the METH is emitted to the atmosphere or thrown into landfills. The bottom line is that virtually all of the METH that is produced and used is eventually released to the atmosphere whether it be in plants during metal cleaning or foam production or by consumers in their use of a product.

MEASURING SOURCE REDUCTION

Many people in the source reduction community are calling for the reporting of rough materials balance information. They believe it would help track trends in hazardous substances use if-throughput information could be collected in addition to the Title III release data. They claim that collecting two additional pieces of data, the first and last lines of Table 1, the amount of the substance brought on site and the amount produced or sent off site in the product, would allow a throughput analysis that could be used to measure progress in source reduction.

Now that we have set the stage with the four plants, let us consider how we might measure progress in source reduction if we could have all the data in Table I reported. There appears to be a consensus that source reduction applies to all media--air, land and water. As mentioned earlier, there is no clear consensus on what source reduction actually encompasses--whether it is restricted to in-process recycling or whether on- or off-site recycling and on-site treatment should be included.

Ignoring this issue for the moment, we need to decide how to measure progress in source reduction if we can agree on what it is and this is not a simple undertaking. There are changes in economic activity from year to year and many people have suggested that to normalize out the effects of growth or a decline in economic activity, we must measure source reduction from year to year on a per unit of product output basis. In plant #1, for instance, if air, waste and water releases declined from one year to the next and the production level of METH remained the same or increased, then we could agree that source reduction had been accomplished. In effect, the total releases per pound of METH produced declined, signalling progress in source reduction.

It is much more difficult in the case of the other plants in Table 1 to normalize according to the weight of product produced. The light fixture manufacturer may change his product mix from year to year depending on orders. Should the plant report the number of fixtures, the pounds of each type of metal cleaned or the surface area of metal cleaned? The paint formulator faces the same problem. Should he report on the number of units filled even if some types of paint require more METH in the formulation than others? The foam producer could report METH use per pound of foam output. Even here, this is a problem, however, since softer foam requires more auxiliary blowing agent. The mix of foam may also vary from year to year.

A second way of normalizing that has been suggested is to use sales. Perhaps we could measure METH use per sales value of the product. Again, however, product prices vary from year to year and the mix of products will often determine the sales value realized.

Another approach that has been discussed is to simply measure source reduction as a reduction in use or releases from year to year. This technique has problems too, and they arise because there is no clear definition of source reduction. The light fixture manufacturer may put in a carbon absorption unit that captures 50 percent of the METH that was previously emitted to the air. The plant reduces its air releases by 3,125 pounds and reduces purchases by the same amount by reusing the captured METH. However, carbon adsorption does not fit most definitions of source reduction which exclude on-site treatment. Under the strict definition, in this case, the plant has not accomplished source reduction, even though air releases have been cut in half and use of the substance has declined.

The plant currently sends its spent solvent to an off-site recycler, but purchases virgin solvent for use in the cleaning operation. What happens if that plant decides to purchase recycled solvent and use it in place of virgin solvent? All of the values in Table 1 remain the same so that this plant has not accomplished any source reduction. Virgin solvent purchases have declined in the economy, however, so source reduction has in fact been accomplished on a nationwide basis, reducing the requirement for virgin production. Under some definitions there has been no net source reduction because off-site recycling is not considered a permissible source reduction option.

What if the light fixture manufacturer converts to a new heavy hydrocarbon solvent that is not on the Section 313 list. All definitions of source reduction include substitution as a viable option. This heavy hydrocarbon is combustible and poses a danger to workers. It is not exempt as a contributor to photochemical smog, and it is relatively unscrutinized in terms of chronic health effects. In fact, it may ultimately prove to be a carcinogen even though it is not currently on the Section 313 list. Through substitution of this new solvent, has source reduction really been accomplished and if so, how much?

A similar measurement problem arises if the light fixture manufacturer converts to an aqueous cleaning system. The water, after it has been used for cleaning, contains 13 greases, oils and metals from the machining process. If the local sanitation district does not require treatment before release to the sewer, then metals are being placed in the sewer where they would not make their way if METH were still being used. Has source reduction been accomplished and if so, how much? If the sanitation district does require the water to be treated, the light fixture manufacturer might have to install an ion exchange process to treat the metals. Ion exchange is an on-site treatment method that generates a metal sludge requiring disposal. We thus have a situation where a substitution--allowed under the strict definition of source reduction--forces the adoption of on-site treatment, a procedure discouraged by some source reduction advocates. Has source reduction been accomplished and, if so, how much?

A similar problem with chemical substitutes might arise in the case of plant #4. In July,, an EPA regulation will cap the production of the fully halogenated chlorofluorocarbons (CFCS) at 1986 levels. The CFCs-will be phased out altogether over the next decade. Roughly half of the auxiliary blowing agent currently used in flexible slabstock foam production is CFC-11; the other half is METH. In July, after the production of CFC-11 is restricted, its price will rise and some foamers will switch from CFC-11 to METH. Other alternatives are being investigated but they probably will not be available on a wide scale for several years.

Because of the regulation on CFC-11, the foamers are unable to practice source reduction of METH and they will likely increase its use in the next few years. We have a situation where one office of EPA, concerned about damage to the ozone layer in the stratosphere is regulating a chemical--CFC-11--that poses a threat to the ozone layer. This office has imposed a regulation that will increase use of another chemical under scrutiny for various reasons by other offices of EPA.

The most comprehensive method for measuring source reduction has been proposed by the Natural Resources Defense Council (NRDC) (Smith, 1988). The concept of throughput is defined to include the sum of total releases, the amount leaving a plant in or on a product, the change in inventory, the amount transformed on-site, the amount recycled on-site or sent off-site for recycling, and the amount entering all downstream processes. Another term--efficiency--is defined by NRDC as "the ratio of the total amount of each hazardous chemical released annually from the processes at a facility (and from subsequent recycling operations) to the throughput in the same year of that chemical at the facility." Releases are the sum of losses from the manufacturing process prior to treatment, from losses from on-site recycling and losses leaving the facility as impurities in a product. Note that the NRDC definition of efficiency is at odds with the common definition. In the NRDC model, a lower "efficiency" (ratio of releases to throughput) is more desirable than a higher "efficiency".

The NRDC analysis includes an example similar to the light manufacturer of Plant #2 in Table 1. In this case, a facility uses another chlorinated solvent--1,1,1-trichloroethane (TCA)--for degreasing metal parts. Losses include 14,000 pounds of atmospheric emissions and 7,000 pounds of waste spent solvent that is sent off-site to a recycler. There is no solvent sent off-site in the product, none in inventory, and none converted on-site to another substance and none entering downstream processes. Thus,, the total throughput is 21,000 pounds. Since the facility receives credit for the recycling, the only release is the atmospheric emission. The "efficiency," therefore, is 67 percent, the ratio of the atmospheric loss to the throughput.

In this definitional regime, reducing the "efficiency" (the release) is equivalent to adopting source reduction measures. Again, we can do this in several ways. A carbon adsorption device could capture emitted TCA before it is released to the atmosphere. Under the NRDC definition, carbon adsorption would be defined as a treatment option because the TCA is released from the process (the degreaser). Use of this treatment option does not reduce releases from the process. The facility could substitute a heavy hydrocarbon solvent with unknown health and environmental effects or it might change the process to aqueous cleaning. In these cases, the plant has no releases of a listed chemical and has therefore achieved zero "efficiency"--a perfect score. The problem with these alternatives as discussed earlier, however, is that they might actually increase damage to health and the environment.

Another problem with this approach is that it may not accomplish what is intends. The NRDC model gives a credit for spent solvent sent to an off-site recycler and does not consider it to be released. The recycler, however, may not recycle the material at all. It may end up in an incinerator, it may be illegally disposed of in a landfill, it may be allowed to evaporate, or it may be poured into a water system. None of these destinations is allowed under the NRDC definition of "efficiency".

On balance, no very good way to measure source reduction has yet been proposed and the NRDC model has significant problems. There remain in all schemes, the problem of the definition of source reduction or an equivalent term. The other problem that arises in all cases is the increase in-use or release of alternative chemicals or outputs from process modifications. These may themselves prove to be dangerous but simply in a different way.

This suggests that there is no generic way to measure or accomplish source reduction across or within industries. In fact, we must do the accounting on a case-by-case basis and evaluating whether or not progress in source reduction has been accomplished involves making a value judgement. Each industrial process and, indeed, each plant involved in a particular industrial process, is unique. Source reduction options that are appropriate for one operation may actually endanger workers or increase hazardous materials generation in another plant. Deciding on the best way to protect human health and the environment will require a great deal of study. Understanding industrial processes, their release characteristics and the implications of the feasible source reduction options is essential.

REGULATION OF SOURCE REDUCTION

There is a growing movement that believes industry has been slow to embrace source reduction. This group believes that only through regulation will there be widespread awareness and adoption of source reduction measures. Two types of regulation are being proposed. The first would require all firms falling under the rubric of Section 313 of SARA to report on the source reduction and recycling activity for the chemicals on the 313 list (H.R. 1457, 1989). The second, based on the NRDC model, would require firms to accomplish 95 percent source reduction over the next decade (RCRA Reauthorization Bill, 1988). It would also apply to the chemicals on the 313 list and the firms in the manufacturing sectors specified in the Title III legislation.

There are three types of firms affected by Section 313 of SARA. The first type is the chemical producers which are very familiar with the chemicals that are the target of source reduction because these chemicals are the product they manufacture. Producers generally have made good progress in source reduction; indeed, it is their business to do so. They must continually attempt to further maximize their yield and the way to do that is to minimize releases.

The second type of firm is also large--an electronics manufacturer or an aerospace firm, for instance--but is not familiar with the target chemicals. These firms do not manufacture chemicals but rather use them in the course of making another product. In the last few years, these manufacturers have established large staffs and correspondingly large budgets to focus on Waste minimization because they believe they can reduce costs and minimize liability in doing so. They have thousands of different chemical formulations moving into their plants each year and they are trying to track these and understand their uses.' The large manufacturing firms are not as far ahead as the chemical producers but some of them have made impressive progress and others-will do so in the next few years.

The third type of firm is small or medium sized and, like the large firms, uses chemicals only to make a product. Some of these firms are examining source reduction but many of them have neither the resources nor the expertise to adopt source reduction and this is unlikely to change significantly in the future. In fact, although many of these companies fall under the rubric of Section 313, most of them are probably unaware that it applies to them and it is likely that they have not submitted data to EPA on their releases.

The proponents of the so-called information regulation argue that we need data on what is actually happening out there today and we need to continue to collect data to measure source reduction trends. In fact, there are really only two reasons to collect such data. First, we don't know ourselves what source reduction is or how to measure it and we are hoping that an immense data collection effort will somehow tell us. Rather than doing the work of understand the industrial processes ourselves, we want industry--the only people who really do understand the processes--to tell us. In effect, this is similar to other data collection efforts where we don't know what to ask initially so we cannot analyze the data that are received. Second, we want the data to establish a baseline for a more prescriptive mandatory source reduction regulation.

The large firms described above would probably be able to cope well with a data submission. They already have staffs who fill out the numerous forms required under various regulations. In contrast, a data requirement would impose a significant burden on small and medium sized firms. Some of these firms would likely have to hire consultants to analyze their records or release streams for them because they lack the expertise or the time to do it themselves. Some would not have the resources to hire a consultant and they would go out of business. Still others would probably simply not respond at all.

Title III of SARA actually applies to thousands of small quantity generators in the nation. The statute requires reporting from users of 10,000 pounds annually of a listed chemical. Looking again at Table 1, a metal fabricator like plant #2 using 10,000 pounds of METH would probably emit three-fourths of it, or 7,500 pounds to the air. The balance, or 2,500 pounds would be hazardous waste. We define small quantity generators (SQGs) and very small quantity generators (VSQGs) as firms that generate 2,200 pounds and 220 pounds per month respectively. This translates into 26,400 pounds per year for an SQG and 2,640 pounds per year for a VSQG. If the release characteristics are similar to those of Plant #2, a metal fabricator using 10,000 pounds per year would fall below the cutoff of a VSQG and there are thousands, perhaps tens of thousands of them in the nation. Those who want source reduction reporting would impose a financial burden on these small businesses that could prove intolerable in many cases.

The other proposed regulation would actually require firms to accomplish source reduction. How would we implement such a regulation? Who would decide whether source reduction had been achieved and who would decide whether it amounted to 95 percent as required? In fact, the bill avoids this question altogether by charging the administrator to specify how throughput should be calculated to establish a baseline for the required 95 percent source reduction.

To illustrate that such a regulation would be an undertaking beyond current public sector capabilities, let us once more refer to Table 1. The chemical producer of Plant #1 already has a fairly low release. The plant manager has worked hard to minimize the losses. The firm already recycles still bottoms back into the reactor and the waste cannot be reduced further with current technology. All the valves and pipe fittings have been checked and those that were leaking have been replaced and a regular maintenance program has been initiated to inspect for leakage. The "efficiency" according to the NRDC model of this product is currently 7 percent. To meet the regulation, the plant will have to reduce the "efficiency" further, to 5 percent. Is it fair to ask the chemical producer who has already instituted source reduction measures to reduce the releases further? It may prove technically impossible to do SO.

It is even more unfair to ask facilities that use chemicals dispersively to achieve a 5 percent "efficiency." A case in point is the light fixture manufacturer who used METH in the course of making a product. The plant's current "efficiency" is 75 percent, significantly higher than the 5 percent mandated by the regulation. The plant presently sends the spent waste solvent to an off-site recycler but does not buy back or use recycled solvent in the process. If the manufacturer instead decides to recycle the solvent on-site--a source reduction option that allows the plant to reuse the spent solvent--then the plant actually pays a penalty. If we assume that 90 percent of the solvent can be recycled in this process, then "efficiency," in this case increases to 77.5 percent. This occurs because atmospheric losses remain the same but the waste loss increases because the plant generates a still bottom from the distillation that must be sent off-site for disposal. The details of this calculation are described in the Appendix.

What happens if we try to do something about the much larger atmospheric loss? The plant likely has an old vapor degreaser. Replacing it with a new unit with a second set of condensing coils and a higher freeboard might reduce annual atmospheric losses by 30 percent, from 18,750 pounds to 13,125 pounds. If we assume the plant sends its waste to an off-site recycler, then "efficiency" with the new degreaser is 68 percent--still very far from the 5 percent required. Let's assume that instead of buying a new degreaser, the plant puts in a carbon adsorption unit. This unit can reduce atmospheric releases by 80 percent, from 18,750 pounds to 3,750 pounds. This lowers the "efficiency" to 15 percent and to lower it further would probably not be technically feasible.

The plant, even to adopt these measures which are insufficient, would incur significant capital costs. Together the still and the carbon adsorption device would require at least a $10,000 investment if steam were already installed. If not, as is generally the case in California, the cost would be much higher (Mooz et al, 1982). Furthermore, under most definitions of source reduction, carbon adsorption, a treatment technology, is not included and it might not be allowed under
the proposed regulation.

The only real alternatives the plant has--and the plant is-after all an SQG--is to switch away from the listed chemical or to go out of business. As discussed earlier, the plant might convert to a combustible heavy hydrocarbon solvent that may eventually cause health or environmental problems. Or it might convert to an aqueous cleaning system which carries metals and organics into the sewer. By definition, in either of these cases, the conversion to a non-listed chemical accomplishes 100 percent source reduction or leads to perfect "efficiency." Unfortunately, we have no way of knowing if the conversion will better protect human health and the environment. Indeed, it may cause other, more dangerous problems than exist currently.

THE REAL ISSUE

What is really driving the movement toward regulation of source reduction? one of the reasons for this intense focus is that pollution of air, water and land is truly serious and we must address our widespread use of hazardous substances. Source reduction--which, in principle, has no opponents--is a vehicle that allows us to question the life cycle production and use of these substances to minimize pollution. Another related reason for imposing a regulation requiring source reduction is that such a regulation may be able to reduce, and indeed eliminate, the hazardous substances that we have not been able to eliminate through other regulatory statutes.

We devote significant resources to regulating hazardous substances once they have entered commerce through a variety of statutes. The Occupational Safety and health Act (OSHA) regulates exposure in the workplace; the Clean Water Act (CWA) and the Clean Air Act (CAA) prevent pollutants from entering the water and air respectively; the Resource Conservation and Recovery Act (RCRA) regulates hazardous waste; The Comprehensive Environmental Response and Compensation Act (CERCLA) governs the cleanup of contaminated sites; and Proposition 65 in California is intended to prevent substances that cause cancer and birth abnormalities from entering drinking water supplies.

We have in-place a set of expensive, frequently ineffective and often inconsistent regulations. They are largely command and control; that is, they prescribe exactly what kind of equipment or procedures to use for specific situations. Because such regulations are generally not cost-effective, they depend on government enforcement to exact compliance. Indeed, compliance with the regulations is likely to be low unless there is a very large inspection staff and the funds for such staffs are rarely available. Furthermore, because of the fragmented statutes, each inspector focuses only on the medium in question. That is, the hazardous waste inspector does not inspect a plant for CAA violations involving atmospheric emissions. Even within a statute, there is conflict; the inspector looking for violations of tropospheric smog regulations will encourage the substitution of stratospheric ozone depleters. A final problem is that some inspectors are not well trained and they may not be knowledgeable.

How has this situation come to pass? The regulatory regime has evolved to its present state because policy makers did not assemble it using a systems approach. The prevailing view, at least implicitly and often explicitly, was that pollution could not be prevented or minimized unless industry was told what to do and how to do it in considerable detail. No one wants carcinogens in the ground water or in their basements. For this reason, there is a huge constituency for regulations like the Clean Air Act and CERLA, which may be termed "back end" regulations. The CWA and RCRA also fall into this category. There is no constituency for hazardous substances once they have been used and no one wants them in the air, land or water.

The emphasis on a regulation requiring source reduction signals the failure to achieve appropriate, efficient and consistent controls on hazardous substances once they have been used. Some proponents believe that, in one stroke, a regulation requiring source reduction will do what all the other regulations together have not been able to do--eliminate or significantly reduce the use of hazardous substances in commerce. The real question that requires societal debate is "do we want to eliminate or significantly reduce the use of hazardous substances in commerce, and if so, how should we best do it?"

There are a number of substances in commerce today that cause cancer in laboratory animals. The Toxic Substances Control Act (TSCA) and the Federal Insecticide, Fungicide and Rodenticide Act (FIFRA) have been largely ineffective in preventing these substances from entering the market in the first place or in taking them off the market once they are there. Once a substance enters commerce,, it is woven within the fabric of society. Workers produce it; it may be used to produce other chemicals; it is used in products or to make products. There is a vested interest in continued use of this chemical because of the revenue it generates for the producer, the jobs it creates, and the convenience it offers to all of us to maintain a good standard of living. In effect, these sunk costs guarantee a large constituency for continued use of the chemical. Once the substance is "spent," however, it no longer has a constituency and herein lies the paradox. We want to use the substance for the benefits it provides but we don't want to pollute our environment with it. Thus we have been led to the inefficient "back end" regulatory regime that exists today.

At this stage, we must admit the error, come to terms with it, try to understand it and open up to societal debate the questions about how to solve it. Do we want human or animal carcinogens on the market? If so, we must focus on the best way to minimize their impact on human health and the environment. There may be efficient ways to do this or there may not. If not, we should concentrate on removing them from the market from the outset through TSCA and FIFRA which were designed exactly for that purpose.

RECOMMENDATIONS

Any kind of field work leads to the unavoidable conclusion that firms need help in managing the hazardous substances they use. The concept of source reduction remains confusing. Mandating source reduction without understanding what it is and without understanding what its consequences are would be a mistake. There is no assurance that a source reduction regulation would lead to better protection of human health and the environment. Indeed, it might end up forcing firms to adopt other products, processes and chemicals that will ultimately prove dangerous in a different manner. We are at a critical juncture now and we have the opportunity to reevaluate toxic chemical regulation under the rubric of source reduction. The incentive we supply in the current regulatory regime is compliance. Because most of the regulations are ill conceived and in conflict with one another, compliance may not lead to better management of hazardous substances. We should consider carefully what the best course is, not to exact "compliance" through another command and control regulation but to better ensure environmental and human health protection.

I have personally become convinced that command and control regulations on source reduction now would be a mistake. Our top priority now should be to provide technical assistance to small and medium sized plants on an industry-by-industry, plant-by-plant basis. Although a component of the assistance would obviously be source reduction, the major effort would focus on better hazardous substances management. We should take the next few years to try to understand what source reduction is and the only way of doing this is to learn about the industrial processes in-depth. This will require a significant time investment, an open mind and considerable dedication.

Small and medium sized firms are not the only ones who would benefit from a concerted technical assistance program. As the technical assistance teams increased their knowledge of particular industries, they could also provide a valuable service to large firms who use hazardous substances. Although such firms have technically trained staffs devoted to waste minimization, knowledgeable outsiders can facilitate better hazardous substances management in a variety of ways. They can act as ombudsmen between firms offering alternative chemicals, products and processes and firms looking for alternatives. Because of their knowledge of processes, they can critically evaluate the alternatives and try to understand their consequences. They can use informed judgement to suggest promising alternatives. They can hold meetings on specific processes to identify common problems and common solutions. Finally, they can document the successes and failures of various options for widespread dissemination.

Everyone wants to play a role in spurring society toward source reduction. It is the seminal issue of the decade. We must take care that these efforts do not waste resources and that they do not make things worse than they are. It is the responsibility of the source reduction community to learn what source reduction really is before mandating that it occur. It is the responsibility of the source reduction community to learn about the complexities of chemical use and about the industrial processes that employ them. It is the responsibility of the source reduction community to understand the tradeoffs involved in the adoption of source reduction measures. It is the responsibility of the source reduction community to examine the issue of how to-better protect human health and the environment.

APPENDIX

"EFFICIENCY" CALCULATION FOR THE METAL FIXTURE MANUFACTURER

The NRDC model contains two concepts. The first is throughput which is defined as the sum of:

- total releases,
- the amount leaving a plant in the product,
- the inventory change,
- the amount recycled on-site or sent off-site for recycling,
- the amount entering-all- downstream processes.

The second concept is "efficiency," which is defined as the ratio of releases to the throughput.

The light fixture manufacturer in Table 1 in the text purchases 25,000 pounds of METH annually. The losses are 18,750 pounds of atmospheric-emissions and 6,250 pounds of spent waste solvent. This firm currently sends the waste solvent off-site to a recycler. The "efficiency" is therefore 75 percent. That is the releases are 18,750 pounds and the throughput is 25,000 pounds.

USE OF OFF-SITE RECYCLED SOLVENT

The plant sends its contaminated solvent off-site to a recycler but buys virgin solvent for use in the process. If the plant decides to instead purchase recycled solvent, the "efficiency" would remain the same as above--75 percent. The firm does not receive a credit for converting from the use of virgin solvent to the use of recycled solvent. Indeed, according to the NRDC model, the plant does not even have to use recycled solvent to get a benefit for sending solvent to a reclaimer. As mentioned in the text, there is no way of knowing the ultimate destination of the solvent. This is a failing in the model.

ON-SITE RECYCLING

What if the plant decides to purchase a still for recycling on-site? If we assume the still to be 90 percent efficient, then out of the 6,250 pounds of spent solvent, 5,625 pounds goes back into the-process and 625 pounds is sent off-site for disposal. In this case, releases are the sum of the atmospheric loss (18,750 pounds) and the waste loss (625 pounds) or 19,375 pounds. The throughput is the sum of the releases (19,375 pounds) and the amount recycled on-site (5,625 pounds) or 25,000 pounds. "Efficiency," in this case, is 77.5 percent.

This value is higher than the "efficiency" of off-site recycling, an anomaly of the model. Indeed, a regulation patterned on this model would provide a disincentive to adopt on-site over off-site recycling, an outcome that is probably not desirable.

PURCHASE NEW DEGREASER

Purchasing a new degreaser would lower the atmospheric losses by 30 percent, from 18,750 pounds to 13,125 pounds. If we assume the plant sends spent solvent to an off-site recycler, the throughput would be the sum of the atmospheric losses (13,125 pounds) and the spent solvent sent off-site (6,250 pounds) or

19,375 pounds. "Efficiency" is the ratio of the releases (13,125 pounds) to the throughput (19,375 pounds), or 68 percent.

PURCHASE CARBON ADSORPTION DEVICE

If the firm purchases a carbon absorption unit instead of a new degreaser, the atmospheric emissions can be reduced by 80 percent, from 18,750 pounds to 3,750 pounds. In this case, releases are the remaining atmospheric emissions (3,750 pounds). Assuming the adsorbed METE is desorbed and reused, then throughput is the sum of the releases (3,750 pounds), the amount sent off-site for recycling (7,000 pounds) and the amount reused on-site (15,000 pounds) or 25,000 pounds. "Efficiency" is therefore 15 percent.

Note here that carbon adsorption is extremely expensive and most plants would be unable to purchase it. It also takes significant expertise to operate. Furthermore, it is not clear that this method--which, in fact, is a treatment technology--would be allowed under the regulation.

SUMMARY OF SOURCE REDUCTION OPTIONS

It is doubtful that plants could reduce their atmospheric losses further than with the carbon adsorption device. The more cost-effective option would be to convert to another solvent not on the Section 313 list or to an aqueous cleaning system. As discussed in the text, this conversion may pose problems but simply in a different way from METH.

This example raises another point as well. After the carbon adsorption device has been purchased, virgin purchases for the firm amount to 10,000 pounds annually. This includes the waste loss (6,250 pounds) and the atmospheric loss (3,750 pounds). Although the plant actually uses 25,000 pounds of solvent, the 15,000 pounds of solvent captured by the carbon adsorption device is reused in the cleaning process and substitutes for virgin solvent. Section 313 of SARA requires chemical users of 10,000 pounds annually to report. If the light fixture manufacturer lowered "use" one pound further, the plant would escape the reporting regime altogether and would no longer have to report. Because the regulation requiring "efficiency" of 5 percent in ten years is patterned on Section 313, this plant would escape the requirement after achieving an "efficiency" of 15 percent--not 5 percent. This is an anomaly of the regulation.

REFERENCES

Federal Register, "Toxic Chemical Release Reporting: Community Right-to-Know; Final Rule," 40 CFR Part 372, February 16, 1988, p. 4500.

H.R. 1457, March 15, 1989.

RCRA Reauthorization Bill, September 9, 1988.

Smith, Ned Clarence, "The Use of Mass Balance Data in the Natural Resources Defense Council's Proposed Model Waste Reduction Program," March 1988.

Wolf, Katy, "Source Reduction and the Waste Minimization Hierarchy," Journal of the Air Pollution Control Association, Vol. 38, #5, p. 681, May, 1988.

Appendix B

Measuring Change in Ecosystems: Research and Monitoring Strategies

A Workshop Report

Summary

Scientists began sustained measurements of ecosystems several decades ago. Classic studies, such as those of Hubbard Brook in New Hampshire, and recent studies, such as that of Chesapeake Bay, demonstrate the evolution of ecosystem science. These studies have heightened scientists' awareness of the need to understand better the continued degradation of many of the country's ecosystems, particularly where the causes of such degradation are anthropogenic.

The Science Advisory Board of the Environmental Protection Agency (EPA) noted this need in its subcommittee report, *Future Risk: Research Strategies for the 1990's* (EPA, 1988). An increasing number of resource agencies also recognized the need to develop programs for evaluating the consequences of their management decisions and the public's use of the ecosystems under their control. Unfortunately, as indicated in the 1977 National Research Council (NRC) report, *Environmental Impacts of Resource Management*, efforts among government and private institutions to develop programs and methodologies to address ecosystem degradation have suffered from lack of top-level support and coordination.

An NRC workshop was held in Warrenton, Virginia, on March 2-3, 1989, to develop ideas for research that EPA and other agencies could use to address ecosystem and landscape issues. Through this report of the workshop, the workshop participants, while recognizing problems inherent in dealing with the wide diversity of systems found in this country, hope that their efforts will encourage stronger coordination among agencies.

According to workshop participants, this absence of top-level support and coordination continues today while our nation's natural systems continue to deteriorate. New approaches, such as ecological risk assessment, have developed over the past decade. These approaches, combined with increasing effort in ecosystem studies, may give us the necessary tools to retard the rapid degradation of many of our nation's natural treasures as well as the systems on which we depend for our well being.

To improve the nation's ability to anticipate future ecosystem and landscape changes in response to natural and anthropogenic perturbations, workshop participants developed a strategy that included the following principal elements:

1. Establish the principle of maintaining ecosystem and landscape integrity and sustainability as an integrative management policy.
2. Select indicator variables of ecosystem integrity and sustainability on a regional or landscape basis.
3. Develop standard methods of data collection and analysis for ecosystem monitoring.
4. Establish an integrated, large-scale, long-term national program for regionally focused ecosystem monitoring, research, and risk assessment.

A program that included these four aspects could assist in solving regional issues through the use of regional experts; coordinate regional and local monitoring programs and synthesize data collected from these programs; rank research activities necessary to provide tools for the national program; provide information for management and decision making; and, through annual reports, inform the public about short- and long-term environmental changes, assessment needs, and management recommendations.

Measuring Change in Ecosystems: Research and Monitoring Strategies

INTRODUCTION

The nation is facing concerns about many environmental problems that involve long-term, large-scale environmental degradation (e.g., complex chemical pollution, regional air pollution, coastal degradation, wetland losses, and loss of biotic diversity). Moreover, major ecosystems such as the Chesapeake Bay, Puget Sound, the Great Lakes, the Grand Canyon, and the Great Smoky Mountains are subject to multiple anthropogenic disturbances. Effective conservation and management of these systems will require identification of resources in jeopardy as well as knowledge of the causes of degradation. Assessing and managing risks of this magnitude require an effective strategy for environmental research, monitoring, and assessment.

A workshop on ecosystem risk assessment and monitoring was held March 2-3, 1989, near Warrenton, Virginia, to address these research and monitoring needs. This report presents ideas developed at the workshop on ecosystem risk assessment and monitoring for the purpose of understanding ecosystem change or degradation. Strategies to implement a research and monitoring program in a consistent, cost-effective, and scientifically credible manner were addressed.

EPA's Environmental Monitoring and Assessment Program (EMAP) was being developed when the workshop was held. Several of the workshop participants had a role in developing EMAP, and the similarity of EMAP as currently described by EPA to many of the workshop's recommendations reflects that dual role.

Several workshop participants indicated that successful environmental protection will require greater efforts by federal agencies to understand and thus to predict ecosystem- and landscape-level consequences of environmental disturbance and contamination. The federal government's approach to environmental threats has been focused largely on risks to human health, endangered species, and a few critical habitats. This approach is insufficient, although significant progress has been made in identifying and controlling some major air and water pollutant sources.

Assessments of the response of ecosystems and landscapes (large spatial units with interacting ecosystems) to anthropogenic perturbations are performed in a piecemeal, fragmented manner because responsibilities are divided among many agencies; such assessments usually fail to provide adequate estimates of environmental threats according to a number of workshop participants. To address inadequacies, a coordinated research program is needed that is directed toward identifying key indicators of ecosystem integrity (structure, function, and stability) and toward using these indicators to provide information about the natural variability of ecosystems. Such a research program is vital for improving our knowledge of the kinds of perturbations at the ecosystem and landscape levels that are likely to be significant risks to the stability or long-term viability of a variety of terrestrial and aquatic ecosystems.

A number of workshop participants indicated that subtle ecosystem perturbations may have long-term regional and even global consequences. Various attempts have been made to improve understanding of the structure and function of large ecosystems, such as the International Biological Program, the International Geosphere-Biosphere Program, the Long-Term Ecological Research Program, and the Global Emissions

Monitoring System. In addition, NRC studies (1981, 1986) have identified the need for more baseline data on ecosystems and their natural variability to improve the ability to predict responses to natural and anthropogenic perturbations.

Public and national concern over the possibility of regional and global environmental consequences of such perturbations has encouraged ecologists and resource managers to increase their efforts to understand the dynamics of ecosystems and landscapes and to predict with greater reliability ecosystem responses to natural and anthropogenic perturbations. Examples of studies of natural resource perturbations at the landscape level are in the literature (Barrett, 1985) and were also presented at the workshop (Patten, 1989). Unfortunately, the available information on ecosystems, including their natural variability, continues to be insufficient to permit rigorous risk assessments, especially at the landscape level. According to a number of workshop participants, recommendations for action to reduce risks to ecosystem integrity presuppose knowledge sufficient to distinguish changes that would have serious adverse consequences from changes that would be either beneficial or insignificant. It is precisely this kind of knowledge that is still inadequate or improperly synthesized.

This workshop was in large part an effort to address this lack of synthesized information for ecosystem research and monitoring strategies. Workshop participants attempted to develop a general research strategy that would lead to better understanding of ecosystem and landscape processes relevant to ecosystem risk assessment and would address long-term environmental issues that are likely to confront the nation over the next decade or more.

PRINCIPAL FINDINGS

During the 2-day working group and plenary discussions, workshop participants indicated that regulatory and resource management agencies should consider the following strategy to improve the nation's understanding of landscape changes resulting from anthropogenic and natural perturbations.

1. Establish the principle of maintaining ecosystem and landscape integrity and sustainability as an integrated management policy.
2. Select indicators of ecosystem integrity and sustainability on a regional or landscape basis.
3. Develop standard methods of data collection and analysis for ecosystem monitoring.
4. Establish an integrated, large-scale, long-term national program for regionally focused ecosystem monitoring, research, and risk assessment.

The elements of each of these is described further below:

1. Establish the principle of maintaining ecosystem and landscape integrity and sustainability as an integrated management policy.

Management of ecosystems and landscapes to prevent continued deterioration of natural resources requires a research program dedicated to that objective according to a number of workshop participants. Sustainability requires maintenance of essential ecosystem resources and processes (e.g., available moisture and nutrients and productivity) above threshold levels determined to be necessary to maintain essential system integrity and vitality. An integrated national monitoring and research program should provide periodic information, based on the analysis of critical indicators, to assess whether this objective is being achieved.

To help provide direction for implementing this suggestion, several workshop participants developed a strategy for assessing and managing risks of large-scale ecological change so that decisions regarding choices among regulatory action, additional monitoring and research, or no action could be made within a risk assessment framework analogous to that now used to make other regulatory decisions (such as decreasing the speed limit on highways or banning use of certain chemicals as food additives).

The key component of the assessment strategy is an ecological inventory, monitoring, and research program to characterize the "health," or integrity, of the environment (Shaeffer et al., 1988) and to quantify the influence of potential environmental perturbations. The data from this program, together with appropriate statistical and mechanistic models, would be used to (1) identify potential adverse changes before major or irreversible damage has occurred, (2) identify causes for the observed changes, and (3) predict ultimate consequences—leading to a rational basis for management decisions (Figure 1).

Selection of end points is the first step in any risk assessment. In health risk assessment terminology, an end point is most often defined as an undesirable event (e.g., contracting cancer or being injured in an automobile accident), and the objective of the assessment is to quantify the risk of occurrence of this event. The appropriate end points for human-health risk assessment are usually negative and obvious (e.g., mortality, morbidity, and teratogenesis). Some negative population-level end points for ecological risk assessment, such as extinction of endangered species or reductions in fish or timber yield, can be readily defined (Barnthouse et al., 1988). However, knowledge of population-level end points is not sufficient to make adequately informed environmental decisions when dealing with ecosystem and landscape perturbations; knowledge of intrinsic variables that indicate changes in ecosystem structure, function, or stability is required.

According to several workshop participants, some critical attributes are difficult to measure on ecosystem and landscape levels, aside from using aerial photographs or satellite observations. Therefore, to perform a risk assessment, changes in those attributes usually will need to be inferred from more readily measured variables. For each critical attribute identified for ecosystems, one or more measurement variables are needed. These variables should be readily measurable, sensitive to different types of perturbation, indicative of current status, and indicative of short- and long-term changes. Suter (1989) defined "measurement end points" as the measurements from which changes in assessment end points are inferred. For example, maintenance of lake trophic status is a common assessment end point (indicator variable) for regulation of discharges into a lake. However, regulation is not usually based on measurements of effects of effluents on receiving systems, such as changes in the trophic status. Instead, regulation is based on measurement of nutrient loading and published water quality standards.

The second step is establishment of a monitoring program to evaluate the integrity of ecosystems or landscapes through their critical attributes. Such a program should provide periodic quantitative information on the measurement variables (e.g., density of critical resource species) and other diagnostic variables (e.g., contaminant distributions, climatic data, and land-use characteristics), thereby assessing the condition of the ecosystem. These measurement variables would be chosen to be region-specific (regional indicator variables).

Third, explicit statistical or mechanistic models are needed to predict changes in the critical attributes based on measured changes in the variables. Statistical extrapolation models, simulation models, or other quantitative risk assessment models could be used to translate variable measurements into estimated values or changes in values of the critical attributes. Measurements of diagnostic variables would help to elucidate the causes of change. Because of the spatial variability in ecosystems and potential multiple cumulative effects, geographic information systems may be crucial to accurate prediction of the effects of different threats on regional landscapes (Brown and Norris, 1988). Management decisions of regulatory action, additional monitoring and research, or no action could then be made within a risk assessment framework analogous to that now used to make other regulatory decisions.

A number of workshop participants also indicated that the assessment and evaluation of critical attributes of ecosystems, analyzed within system boundaries (e.g., changes in biodiversity or rates of nutrient flux) and between ecosystems (e.g., rates of genetic dispersal or of nutrient exchange between systems), should clarify ecosystem self-regulatory and recovery processes and, consequently, suggest strategies for ecosystem management and land-use planning on a regional basis. In regional land-use planning, retention and fluxes between various ecosystems (e.g., between forest and agricultural plots) can be properly evaluated mainly at the landscape level. Changes in regional land-use patterns likely will alter individual ecosystem dynamics within the landscape mosaic. Further, several workshop participants indicated a landscape perspective for ecosystem analysis may also increase our

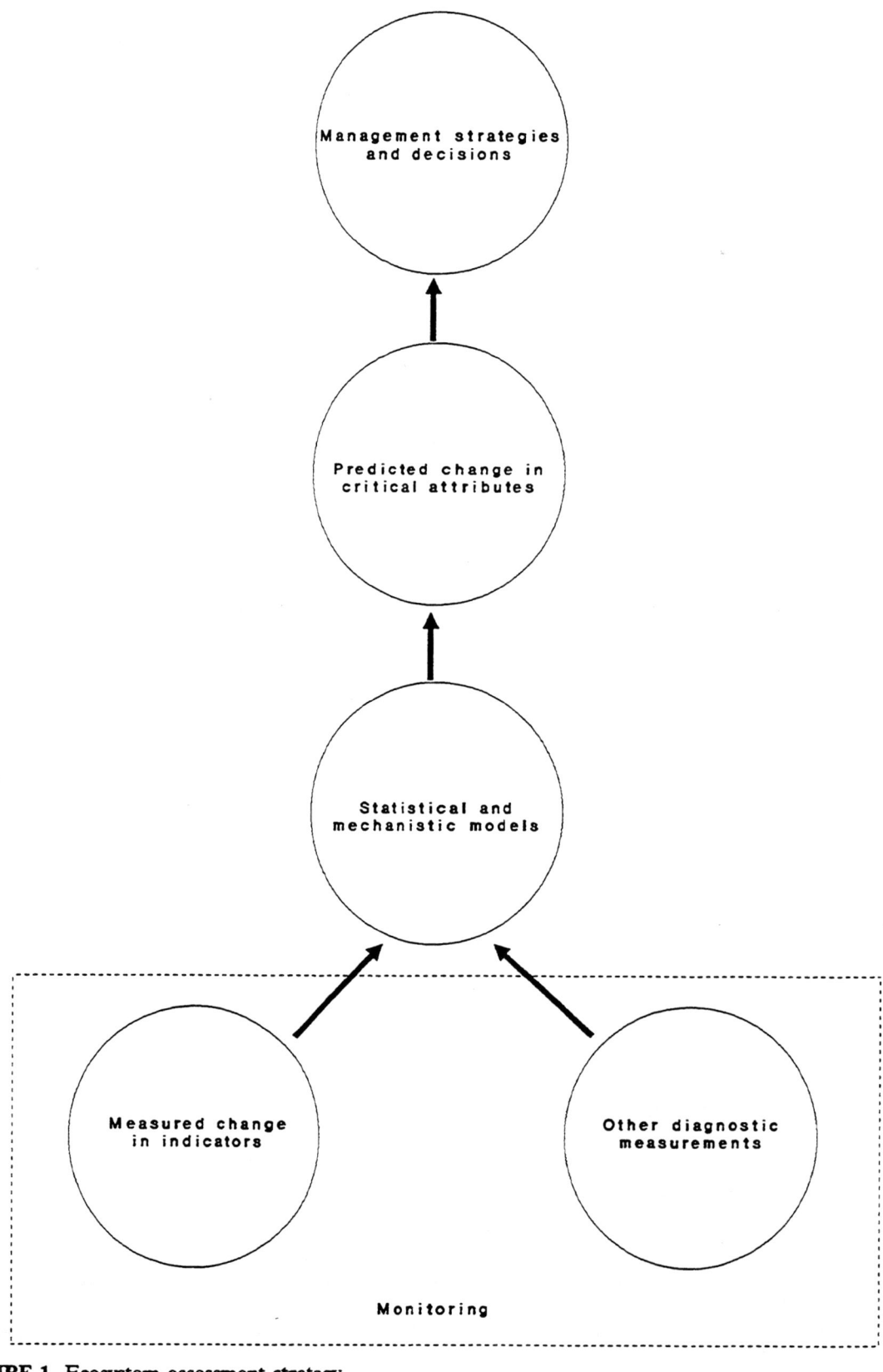

FIGURE 1 Ecosystem assessment strategy

understanding of life cycles of keystone species that may use more than one ecosystem type during their life history. For example, many bird species use natural ecosystems for nesting but use manipulated systems (e.g., agricultural fields or controlled-flow rivers) for foraging. This information should also help to explain changes in biotic diversity within various ecosystems. It may well be that better understanding of the interactions among ecosystems will prove to be more important for risk assessments and ecosystem regulation than the more isolated processes are that ecologists now commonly study, e.g., individual species indices. A research and monitoring program that assesses interactions among systems will require interdisciplinary coordination and integration of research (Barrett, 1984).

2. Select indicators of ecosystem integrity and sustainability on a regional or landscape basis.

Ecosystems consist of the interacting biotic and abiotic elements of a defined physical environment (Odum, 1971; Ricklefs, 1976). Ecosystems can be designated by their dominant physical and chemical characteristics, such as glacial lakes or hardwater streams, or by dominant biological populations, such as oak-hickory forests or big bluestem grasslands.

A number of workshop participants indicated it is impossible to measure all biotic and abiotic elements of ecosystems for changes in structure or function. The alternative approach suggested by several workshop participants is to define characteristics of ecosystems that are likely to be critical for maintenance of ecosystem integrity and sustainability. These characteristics are called critical attributes in this report. Although critical attributes are only components of ecosystems and landscapes, their status can be used to evaluate the linkages that make an ecosystem function as a whole. These attributes are often complex and difficult to measure, e.g., maintenance of a lake's trophic status; therefore, for each critical attribute, one measurement variable or more are needed that can be measured or monitored during long-term studies of ecosystem health and sustainability. For each measurement variable, indicator variables, such as selected species, should be chosen regionally because of the complexity and variability of ecosystems throughout the country. Table 1 shows examples of critical attributes likely to be important in assessing ecosystem integrity.

Because indicators are specific to ecosystem type and will differ among geographic regions, they must be selected on a regional basis by scientists who best understand the local ecosystems. Regional workshops could be held for scientists to discuss and select these indicators. It is imperative that ecologists most knowledgeable about specific regional ecosystems (whether from academia, the private sector, or state and national environmental centers) be involved actively in the selection of the indicator variables to be studied (e.g., controlling nutrients, critical species, critical trophic-level components for key systems, and species indicative of disturbance).

3. Develop standard methods of data collection and analysis for ecosystem monitoring.

Three stages in the acquisition of information needed to assess changes in critical attributes of ecosystems are the following:

• Inventory and compilation of existing information characterizing important features, natural resources, and other environmental baseline data on landscape units or their component ecosystems. This information may be available from a variety of federal and state monitoring programs, natural resource data sets, and literature.

• Regional or national monitoring programs that will provide information to determine the status of critical attributes, to detect change, and to evaluate the effectiveness of selected critical attributes as diagnostic factors. The focus of research programs should be to develop a set of monitoring methodologies that would derive a set of environmental indices that could be used by decision makers (Boyle, 1987, 1989). Monitoring programs may indicate ecosystem changes; however, research is needed to explain the cause of these changes and to interpret the significance and implications of these changes.

• Research at the ecosystem and landscape levels that will provide information on critical attributes that are poorly understood, on the

TABLE 1. Ecosystem Integrity Attribute Assessment Examples

- **Critical attribute:** *Elemental dynamics*—inputs, internal transport, and losses of critical plants, animals, and elemental biochemical compounds, including the flux of nutrients essential for primary production and secondary metabolism.

    Measurement variable: Potentially limiting nutrients, carbon flux, biomass.

    Regional indicator variable: One or more controlling nutrients in regional ecosystems.

- **Critical attribute:** *Energy dynamics (physical)*—energy exchange at geological and biological surfaces (e.g., insolation, sensible and latent heat fluxes, and transportation) and mixing processes (e.g., turbulence, convection, and advection).

    Measurement variable: Microclimatic and hydrogeological processes.

    Regional indicator variable: Specific microclimatic and hydrogeological processes shown to be controlling variables for selected regional systems.

- **Critical attribute:** *Food web (trophic dynamics)*—the set of trophic relationships among species in a community. In its simplest form, the food web is an energy-flow diagram connecting each consumer to all species that it consumes. However, in its dynamic form, the concept also includes rates of consumption, preference of food items, and prey switching. A given food web may indicate which species are necessary resources for other species (e.g., a particular valued species), the amount of redundancy in community functions, and the degree to which particular consumers, termed "keystone" species, may control the competitive processes among the species consumed.

    Measurement variable: Structural characteristics: species density, biomass, and richness; community composition.

    Regional indicator variable: Presence and abundance of species shown to be controlling trophic dynamics in selected regional systems.

- **Critical attribute:** *Biodiversity*—the number of taxa per unit area as represented by populations, guilds, or life forms, as well as the relative abundance of the various taxa.

    Measurement variable: Species richness.

    Regional indicator variable: Populations, guilds, or other selected taxonomic groups.

- **Critical attribute:** *Critical species*—keystone, resource, and endangered species: keystone species are those that exert influences over other populations in their ecosystem out of proportion to their abundances (NRC, 1986); resource species are species of energetic, economic, or aesthetic importance; endangered species are species in imminent danger of becoming extinct.

    Measurement variable: Population monitoring of keystone and resource species

    Regional indicator variable: Species presence, absence, or density.

TABLE 1. Continued

- **Critical attribute:** *Genetic diversity (within critical species)*—genetic diversity represents the number and frequency of different genotypes within species.

   Measurement variable: Biochemical markers.

   Regional indicator variables: Selected species.

- **Critical attribute:** *Dispersal and migration*—movements of individuals within and between ecosystems that are crucial to the population's survival and the ecosystem's health, including colonization or dispersal between habitats and movements of individuals to different habitats for food, reproduction, overwintering, or protection from predators.

   Measurement variable: Dispersal and migration rates.

   Regional indicator variable: Selected species.

- **Critical attribute:** *Natural disturbance*—externally driven disturbances, unrelated to human activities, that have major impacts on ecosystem integrity by altering the species composition, trophic structure, or other important ecosystem developmental processes; these disturbances include wind storms, fires, and floods that are the result of weather patterns and are often essential to maintenance of certain ecosystems.

   Measurement variable: Disturbance events.

   Regional indicator variable: Presence and abundance of invader species known to be indicative of disturbance.

- **Critical attribute:** *Ecosystem development (successional processes)*—developmental changes (successional stages) in species composition through time, mediated by biological-physical interrelationships, resulting in a defined ecosystem structure and function.

   Measurement variable: Structural characteristics: species density, biomass, and richness; community composition; functional characteristics: production/respiration, production/biomass, carbon or nitrogen flux.

   Regional indicator variable: Distribution and abundance of plant and animal communities indicative of successional stages versus those characteristic of mature or stable ecosystems (e.g., r-selected versus K-selected species).

integrity of the ecosystem or landscape, and on the detection and diagnosis of change. Although ecosystem research continues to expand (e.g., the National Science Foundation regularly adds sites to its Long-Term Ecological Research Program), data on specific ecosystem components and their functional roles are always limited. Generation of this type of information, especially as it relates to regional variables and their appropriateness to be used as indicators, must be developed in a research program designed to assess ecosystem integrity in response to natural or anthropogenic disturbances.

According to several workshop participants, specific quality-assurance procedures should also be developed and used by all agencies and institutions engaged in ecosystem monitoring with sufficient rigor to ensure that measurement of physical, chemical, and biological ecosystem indicators are comparable among different regions and over time. These procedures should include adequate taxonomic resolution by experts to ensure accuracy in defining biological indicators. Statistical procedures should be developed to quantify and analyze community and ecosystem-level attributes.

4. Establish an integrated, large-scale, long-term national program for regionally focused ecosystem monitoring, research, and risk assessment.

A regional ecosystem and landscape assessment program is needed to assess the potential large-scale effects of environmental stresses, such as regional air pollution, coastal degradation, changes in land use, loss of biodiversity, and climate change. Information produced by the program could be used to identify incipient adverse changes before irreversible damage has occurred, to determine causes of observed degradation, and to facilitate rational management of ecosystems on a regional level by providing positive feedback to decisionmakers on the consequences of their decisions. Such considerations, indeed, were part of the motivation for developing EMAP (Hunsaker and Carpenter, 1990).

In addition, this program could be designed to achieve several additional objectives: to assist in solving regional issues by providing access to experts; to coordinate regional and local monitoring programs and synthesize data collected from these programs; to rank research activities necessary to provide tools for the national program; to provide information for management and decision making; and, through annual reports, to inform the public about short- and long-term environmental changes, assessment needs, and management recommendations.

After attempting to choose universal measurement variables, several workshop participants realized that the complexity and variability of ecosystems throughout the country are too great to do so. They also realized that they could not develop specific regional research or monitoring projects in this document. They therefore agreed that any organization developing a national program should establish regional working groups of appropriate experts to select the relevant indicator variables for each geographic region. However, workshop participants recognized that some environmental problems also need to be addressed on a global basis. For such problems, large-scale indicator variables such as atmospheric or oceanic components should be selected for monitoring.

This regional approach to ecosystem monitoring and research is not limited to broad national regions such as the northeast or southwest United States, but can also be used at state or local levels. However, a number of workshop participants emphasized that for each geographical region or locale, specific selection should be made of key species or processes that can be used as indicators of changes in ecosystem integrity and sustainability.

Selection of communities and variables to be monitored must also be made relative to expected temporal and spatial perturbations. For example, response to short-term or acute perturbations may be measured best by changes in short-lived organisms (e.g., microorganisms or annual plants), while long-term or chronic perturbations may require monitoring of long-lived organisms (e.g., trees). Different ecosystem processes may also have to be selected as indicator variables of different perturbations. These are decisions that should be made primarily on a local or regional basis.

Expertise needed to address regional or local environmental problems is diverse and often difficult to find. A centralized data base, such as that being developed by the Ecological Society of

America, could maintain a network listing of experts for use by regional agencies or organizations. A coordinated program should be cost-effective and should eliminate many of the inconsistencies and cross-purpose regulations and responses that result from the existing multiagency approach. A coordinated program should not, however, create uniform national management strategies that might not be appropriate regionally.

Although management of ecosystem integrity and sustainability must be considered on a local or regional basis, coordination of these efforts should be integrated. Coordination would ensure the use of compatible methodologies, thus permitting sharing and comparison of information among regulatory agencies and enhancing consistency in evaluating the impacts of environmental perturbations.

CONCLUSIONS

Scientists have conducted ecosystem studies for several decades. These studies have heightened scientists' awareness of the need to understand better the continued degradation of many of the country's ecosystems, particularly where the causes of such degradation are anthropogenic. To improve the nation's ability to anticipate future ecosystem and landscape changes in response to natural and anthropogenic perturbations, workshop participants developed a strategy that included the following principal elements:

1. Establish the principle of maintaining ecosystem and landscape integrity and sustainability as an integrative management policy.
2. Select indicator variables of ecosystem integrity and sustainability on a regional or landscape basis.
3. Develop standard methods of data collection and analysis for ecosystem monitoring.
4. Establish an integrated, large-scale, long-term national program for regionally focused ecosystem monitoring, research, and risk assessment.

A program that included these four aspects could: assist in solving regional issues by providing access to experts; coordinate regional and local monitoring programs and synthesize data collected from these programs; rank research activities necessary to provide tools for the national program; provide information for management and decision making; and, through annual reports inform the public about short- and long-term environmental changes, assessment needs, and management recommendations.

The strategy developed by the workshop participants includes the three key elements of ecological inventory, monitoring, and research to characterize ecosystem integrity (i.e., structure, function and stability). One important part of such a program is the inventory and monitoring of critical attributes, i.e., characteristics of ecosystems that are important determinants of their integrity. These attributes are often complex and difficult to measure, e.g., maintenance of a lake's trophic status or rate of carbon fixation in a tropical forest; therefore, for each critical attribute, one or more measurement variables are needed to describe current ecosystem status and to track short- and long-term changes.

In addition to measurement variables for evaluating ecosystem or landscape integrity (e.g., density of a critical species), statistical or mechanistic models are needed that predict changes in critical attributes from changes in measurement variables. Geographic information systems could aid in evaluating data on ecosystems and potential threats because of their ability to handle spatial information and multiple attributes.

The workshop participants selected some attributes of ecosystems that they considered likely to be important in affecting ecosystem condition (critical attributes). Then, for each critical attribute, variables were selected to be measured or monitored. The critical attributes include the following ecosystem components or processes, together with an example of a measurement variable: elemental dynamics—carbon flux; energy dynamics—hydrogeological processes; trophic dynamics—species density; biodiversity—species richness; critical species—monitoring of keystone species; genetic diversity—biochemical markers; dispersal and migration—dispersal rates; natural disturbance-disturbance events; and ecosystem development-community composition. Examples of regional indicator variables (such as selected species) for each measurement variable are given in the report. These indicators should be chosen regionally because of the complexity and variability of ecosystems throughout the country. Changes in regional land-use patterns should also be evaluated because a landscape perspective in ecosystem analysis will increase our understanding of the interactions within and between ecosystems.

Although this report is organized to recommend strategies on a national level, when attempting to develop specific end points or select indicator species or processes, regional and local planning should be used. Regional working groups of appropriate experts should determine the particular indicator variables for each region. Regional application includes broad regions, such as the northeast United States, as well as state or local areas. Environmental issues on a global basis should also be addressed.

REFERENCES

Barnthouse, L.W., G.W. Suter II, and S.M. Bartell. 1988. Quantifying risks of toxic chemicals to aquatic populations and ecosystems. Chemosphere 17:1487-1492.

Barrett, G.W. 1984. Applied ecology: An integrative paradigm for the 1980s. Environ. Conserv. 11:319-322.

Barrett, G.W. 1985. A problem-solving approach to resource management. BioScience 35:423-427.

Boyle, T.P., ed. 1987. New Approaches to Monitoring Aquatic Ecosystems. Special Technical Publication 940. Philadelphia: American Society for Testing and Materials. 208 pp.

Boyle, T.P. 1989. Selected development needs for assessing ecological risk at the community and ecosystem level. Paper prepared for the National Research Council Workshop on Ecosystem Risk Assessment and Monitoring, Warrenton, Va., March 2-3, 1989.

Brown, N.J., and D.A. Norris. 1988. Early applications of geographical information systems at the Institute of Terrestrial Ecology. Int. J. Geograph. Inform. 2:153-160.

EPA (U.S. Environmental Protection Agency). 1988. Future Risk: Research Strategies for the 1990s, SAB-EC-88-040. Washington, D.C.: Science Advisory Board.

Hunsaker, C.T., and D.E. Carpenter, eds. 1990. Environmental Monitoring and Assessment Program Ecological Indicators. EPA/600/3-90/060. Office of Research and Development. Research Triangle Park, N.C.: U.S. Environmental Protection Agency.

NRC (National Research Council). 1977. Environmental Impacts of Resource Management: Research and Development Needs. Washington, D.C.: National Academy Press.

NRC (National Research Council). 1981. Testing for Effects of Chemicals on Ecosystems. Washington, D.C.: National Academy Press.

NRC (National Research Council). 1986. Ecological Knowledge and Environmental Problem-Solving: Concepts and Case Studies. Washington, D.C.: National Academy Press.

Odum, E.P. 1971. Fundamentals of Ecology, 3rd Ed. Philadelphia: Saunders.

Patten, D.T. 1989. Ecosystem risk assessment: Case studies of natural resource perturbation. Paper prepared for the National Research Council Workshop on Ecosystem Risk Assessment and Monitoring, Warrenton, Va., March 2-3, 1989.

Ricklefs, R.E. 1976. The Economy of Nature. Portland, Ore.: Chiron Press.

Shaeffer, D.J., E.E. Herricks, and H.W. Kerster. 1988. Ecosystem health: I. Measuring ecosystem health. Environ. Manage. 12:445-455.

Suter, G.W., II. 1989. Ecological endpoints. Pp. 2-1-2-28 in Ecological Assessment of Hazardous Waste Sites: A Field and Laboratory Reference Document, EPA/600/2-89/013, W. Warren-Hicks, B.R. Parkhurst, and J.J. Baker, Jr., eds. Corvallis, Or.: U.S. Environmental Protection Agency.

Workshop Participants

Ecosystem Risk Assessment and Monitoring
Warrenton, Virginia
March 2 and 3, 1989

*Duncan T. Patten, Chair, Arizona State University, Tempe
Stanley Auerbach, Oak Ridge National Laboratory, Oak Ridge, Tennessee
*Lawrence W. Barnthouse, Oak Ridge National Laboratory, Oak Ridge, Tennessee
*Gary W. Barrett, Miami University, Oxford, Ohio
Barbara Bedford, Cornell University, Ithaca, New York
Caroline S. Bledsoe, University of Washington, Seattle
*Terence P. Boyle, Colorado State University, Fort Collins
John D. Buffington, Department of the Interior, Washington, D.C.
David Coleman, University of Georgia, Athens
Carl Gerber, Environmental Protection Agency, Washington, D.C.
James R. Gosz, University of New Mexico, Albuquerque
*Clyde Goulden, Academy of Natural Science, Philadelphia
Michael Gough, Resources for the Future, Washington, D.C.
Dale Hattis, Massachusetts Institute of Technology, Cambridge
Edwin E. Herricks, University of Illinois, Urbana
James A. MacMahon, Utah State University, Logan
John Melack, University of California, Santa Barbara
Robert Nisbit, University of California, Santa Barbara
Betty H. Olson, University of California, Irvine
David Parkhurst, Indiana Uiversity, Bloomington
David E. Reichle, Oak Ridge National Laboratory, Oak Ridge, Tennessee
James F. Reynolds, San Diego State University, San Diego
Clifford S. Russell, Vanderbilt University, Nashville
Rosemarie C. Russo, Environmental Protection Agency, Athens, Georgia
Stan Sander, Jet Propulsion Laboratory, Pasadena, California
Thomas Siccama, Yale Forestry School, New Haven
Michael Slimak, Environmental Protection Agency , Washington, D.C.
Lars Soholt, Los Alamos National Laboratory, Los Alamos, New Mexico
Glenn Suter, Oak Ridge National Laboratory, Oak Ridge, Tennessee

* Participants in a follow-up writing session August 23-25, 1989

Workshop Papers

ECOSYSTEM RISK ASSESSMENT:
CASE STUDIES OF NATURAL RESOURCE PERTURBATION

Duncan T. Patten
Center for Environmental Studies
Arizona State University

Paper prepared for the Workshop on Ecosystem Risk Assessment and Monitoring. Committee on Opportunities in Applied Environmental Research and Development, National Academy of Sciences, Airlie House, Warrenton, Virginia, March 2-3, 1989.

Ecosystems are complex systems with populations of organisms interacting with the dynamic fluxes of the system's resources. They are complex to the point that most ecosystem studies emphasize interactions of specific species or flows of selected resources. Studies of the effects of some form of perturbation also tend to be selective; for example, the study of nutrient dynamics of forests that have been altered (Likens et al. 1977).

Ecological risk studies within ecosystems also tend to emphasize the response of (and/or probability of risk to) some component of the system, that component being of interest or importance to society or the researcher. Various aspects of the system may be studied to direct the researcher toward the probability of risk for the species of interest, but the complexity of the interactions within the system are often overlooked.

To extend the more common forms of ecological risk assessment (i.e., those dealing with critical or indicator species) to the ecosystem level, one must be as concerned with exposure to risk of all components and processes within the system as with a single component. Many of these components or processes may seem inconsequential at the time but may be critical to the health of the system once it is fully understood. Thus, ecosystem risk assessment addresses the total system (holistic approach) and is concerned with response of all species and resource flows (e.g., energy, nutrients, etc.).

An apparent contradiction to the holistic viewpoint, however, is the reason for doing ecosystem risk analyses. Although we are interested in the health of the total system, we set priorities on specific resources the system maintains or produces. These resources are the concerns of managers and decision makers who use the probability statements on system or component response to set policy on managing the various resources or allogenic inputs that may alter the system.

A recent review of ecological risk assessment methods (Environmental Protection Agency 1988) discusses various approaches to ecological risk assessment. It compares qualitative and quantitative methods. The former uses professional judgment but may not be useful in setting standards, while the latter uses either quotient or exposure-response methods that allows development of standards for managing risks. The examples of system perturbations mentioned in this review were generally anthropogenic contaminants, the primary concern of the Environmental Protection Agency (EPA).

Ecosystem Perturbation

Ecological risk assessments tend to be oriented toward the effects of contaminants on the system (e.g., Herrick 1987). These are anthropogenic agents that through their properties alter some components or process of the ecosystem. In many cases the ecosystem component is a particular species and the response or fate of that species is the endpoint of a risk assessment. This is not always the case and many other components of the ecosystem may also be examined, especially as receptors of perturbation (Environmental Protection Agency 1988).

Contaminants are one form of anthropogenic perturbations to ecosystems. Other forms include modifications of important resource inputs. Changes in natural resource inputs may have as drastic a series of effects on an ecosystem as input of contaminants, although we tend to be more concerned with the alien, industrially produced products.

Ecological risk studies dealing with contaminants and their effects are of primary importance to EPA; therefore, to gain another perspective, I have selected case studies which include anthropogenic perturbations to ecosystems, but not cases with obvious contaminants. The significance of the selection of cases is a concern for the health of the ecosystem, a concern that needs to be addressed by EPA regardless of the form of perturbation.

Another reason for selecting the cases is that, although the health of the whole system was of concern, there were a broad range of endpoints or resources that were critical for policy setting by the system managers. These resources were not solely biotic but were similar to wildlife, waterfowl, and sediment in the Kesterson Reservoir selenium case (Presser and Ohlendorf 1987), and fish, water, and sediment in pollution studies of the Hudson River (Brown et al. 1985). As human health is the endpoint in risk assessments of various carcinogenic agents, these resources or products of ecosystem processes are the bottomline or endpoint of the risk assessment process in these cases. The use of "bottomline" is appropriate because a negative balance in the welfare of these resources will most often indicate an "unhealthy" or unstable system.

Case Studies

The two cases studies presented in this paper deal with management of the same resource, water. One, the Mono Basin study, deals with limiting water to a natural system, while the other, Glen Canyon Dam study, deals with controlling water flows through a system. In both cases, management of the water resource was for the benefit of large urban areas. The systems differed in that the endpoint resources under consideration by the managers were essentially all nonhuman use resources at Mono Basin, while many resources had potential human uses at Glen Canyon. The systems are simple enough to give a brief overview in this paper.

Mono Basin

An analysis of the Mono Basin ecosystem was published by the National Academy Press following a three-year study (National Research Council 1987). Mono Lake is a closed basin lake on the eastern side of the Sierra Nevadas. It is fed by a series of mountain streams from the Sierras; however, in 1941 four of these streams were diverted as a water source for Los Angeles. The lake level dropped over the next 40 years from an elevation of about 6,420 feet to below 6,380 feet. During this period the lake increased in salinity and the biological balance of the lake and the feeder streams was on the verge of collapse.

The U.S. Forest Service manages the lake and needed to know the effects of the drop in lake level to make lake level recommendations in a management plan for the Mono Basin Scenic Area. This area includes the lake and much of the surrounding area. The study by the NRC committee was not labelled as an ecosystem risk assessment but the approach and interpretations followed that procedure.

A conceptual model of the Mono Basin ecosystem (Fig. 1) shows the relationships between the hydrological cycle, the lake's physical and chemical processes, and the biological components. This model establishes a conceptual organization from abiotic resource inputs through abiotic response and initial biotic response to endpoint resource responses. To better facilitate the initial assessments of the system the model could have been developed in hierarchical scales of organization (Limburg et al. 1986), but the system was understood well enough to make this step unnecessary. The resources (endpoints) of primary concern to the Forest Service are listed along the bottom of the model. These included the birds species, air quality, and shorline vegetation. Also part of the critical resources were the feeder stream riparian systems and the tufa towers, exposed as a result of the dropping lake level but now protected by California law.

The driving variable in this model is the distribution of the water in Grant Lake, that is, either exported or released back into the stream system feeding the lake. The presence or lack of this water alters lake volume which, in turn, influences lake salinity and lake level. These factors then influence other components of the lake ecosystem such as exposed land surface, and brine shrimp and brine fly productivity. The eventual condition of the bird populations, air quality, and tufa result from the relative changes in these factors. Indirectly, one can tie the condition of the resources of concern to the lake volume and water releases into the feeder streams.

As managers of the Mono Basin ecosystem, the Forest Service wanted to know the relative "health" of each critical resource at different lake levels. Each resource responds differently to changes in lake level. For example, dropping lake levels expose more shoreline and may temporarily improve habitat for the snowy plover, while these same conditions increase lake salinity, thus threatening the algae, shrimp, and flies, the foundation of the food chain of the lake. A drop or rise in lake level would also threaten the present condition of the tufa.

The study by the NRC showed gradients in changes for each resource, a form of "exposure-response curve" indicating critical lake levels at which the particular resource was probably at risk of change to levels at which the resource had a certain probability of being lost. The lake levels at which loss began, or loss was certain, varied for each resource (Fig. 2). The health of the ecosystem, i.e., the present condition of biota such as algae and invertebrates (the receptors) and of birds (endpoints), as well as tufa (an endpoint), also varied because the ecosystem had already been perturbed. Thus, ecosystem health becomes a relative condition and future policy based on priorities of endpoint components will establish what level of "health" is desirable or perhaps acceptable. In this case, setting priorities on the related levels of risk allowed for each resource, the Forest Service could establish a range in which the lake level could fluctuate without any long-term harm to major components of the Basin ecosystem.

In establishing its interpretation of response levels of the various ecosystem components, the NRC committee had information available to it from the literature, agency reports, and ongoing research. However, the lake ecosystem was not totally understood. For example, food habits of some of the other bird species weren't well known and reproductive substrates for brine shrimp still were not fully understood. These are examples of processes within this ecosystem that need to be studied to more accurately address the risks to the "critical resources" of altering the lake level.

The analyses in the NRC report were based on the best information available and the expertise of the committee. Some of the response estimates were based on quantitative laboratory and field experimentation, while others were qualitative judgments of members of the committee. Thus, the accuracy of these analyses, as they relate to ecosystem stability, can only be tested by doing the "ultimate" experiment, that is, continuing to lower the lake level. It is inconceivable that this test will be made, although various components of the system need to be monitored closely if the lake is allowed to drop.

Interpretation based on available information and expertise does not guarantee its accuracy. For example, the consequences of raising the level of a subarctic lake were not anticipated in a careful impact assessment, although existing modeling techniques and literature-based information on reservoir dynamics were used (National Research Council 1986). Unfortunately, there was insufficient information on subarctic lakes. This indicates an obvious need for long-term ecosystem studies that would prevent occurrence of similar consequences at other subarctic lake systems under similar or other anthropogenic perturbations.

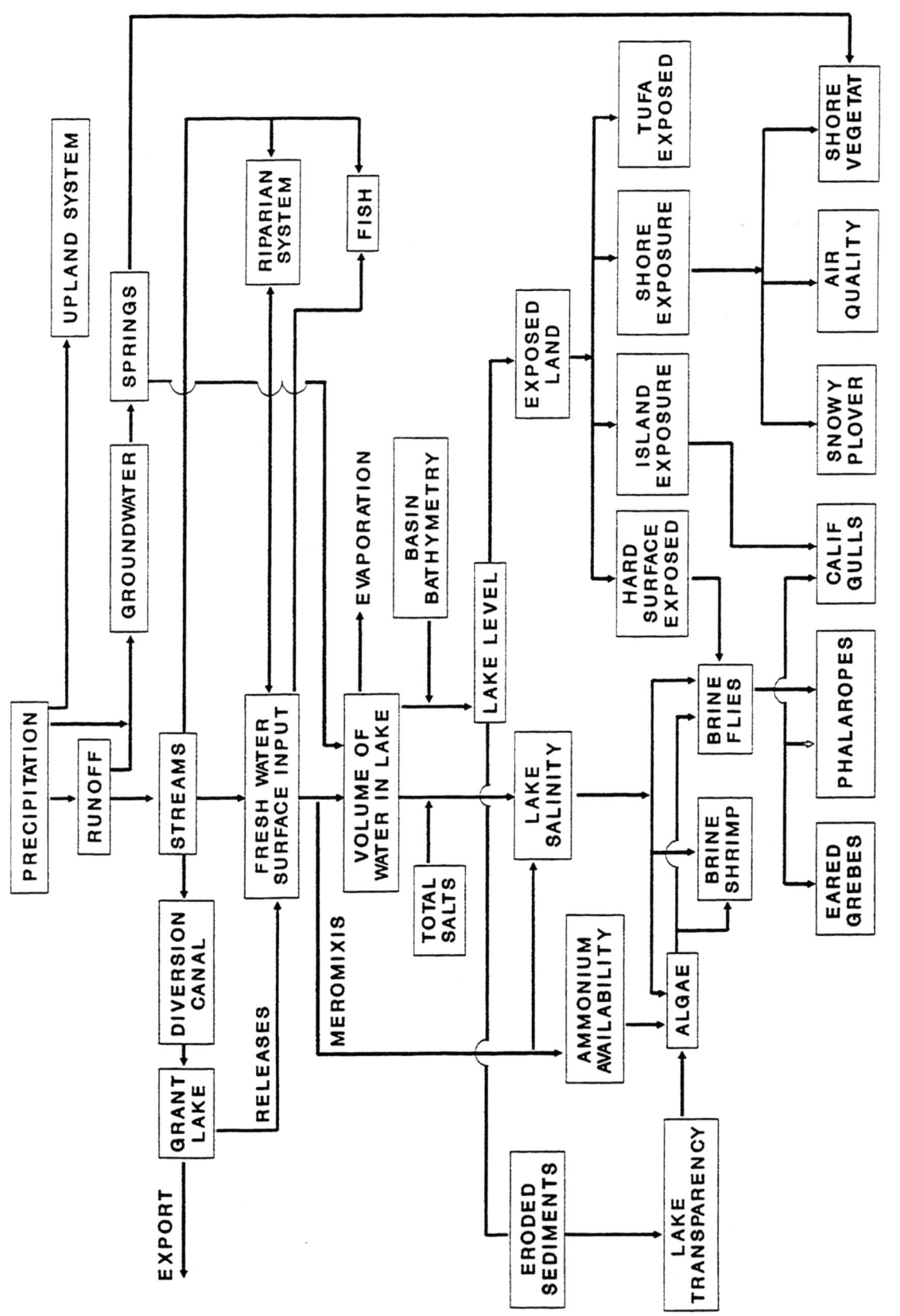

FIGURE 1 Conceptual model of Mono Basin ecosystem

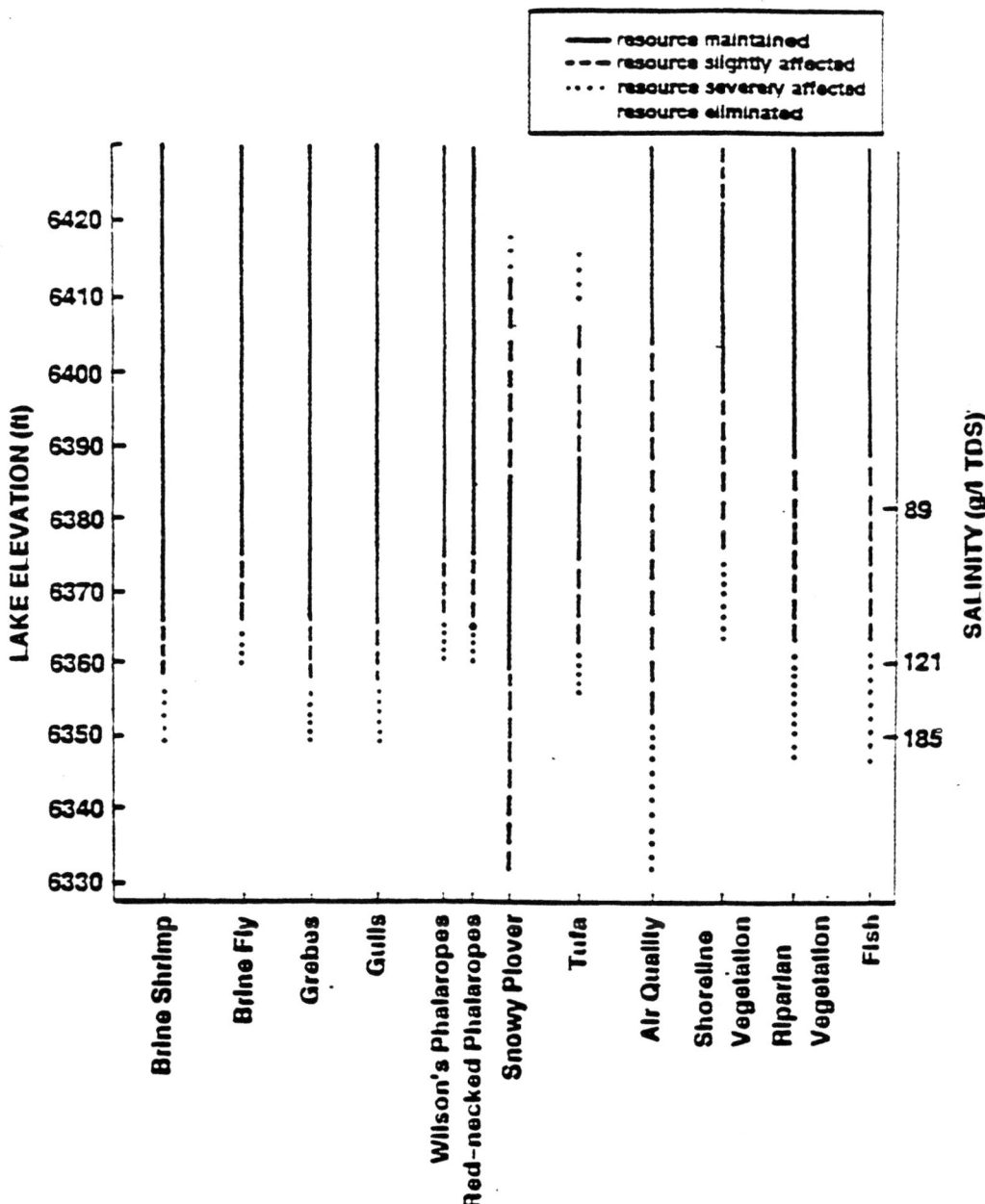

FIGURE 2 Ranges of lake levels affecting resources of the Mono Basin, with three salinities for reference (From National Research Council, 1987)

The gaps in information on the Mono Basin ecosystem also might cause some unanticipated changes if the lake levels were allowed to fall.

Glen Canyon

Glen Canyon Dam on the Colorado River was completed in 1963. It took nearly 20 years for Lake Powell to fill behind the dam. During that period the releases through the dam were based on energy needs of southwestern cities and requirements for water supply to lower basin states. Operation of the dam has been controlled by hydroelectric production with other resources playing a secondary role.

Recent proposed changes in operation of the dam has required a study of the effects of these changes on the many resources of the canyon (U.S. Department of the Interior 1988). Most of the endpoint resources in Glen Canyon and Grand Canyon that would be affected by the releases are directly related to human use of the canyon. These include water for rafting, beaches, maintenance of a trout fishery as well as the general aesthetics of the canyon. Other resources that are of concern, either because of the National Environmental Policy Act or the Endangered Species Act, include bird and wildlife diversity and native fisheries.

An analysis of the effects of changing operations of the dam must take into account the relative risks to the endpoint resources, as well as the receptors, those ecosystem factors that if changed directly or indirectly influence the endpoints (Fig. 3).

Flow control at Glen Canyon Dam (Fig. 3) is the controlling variable for ecosystem responses below the dam. The controlled flows are both in quantity and periodicity. These influence the scouring effects, sediment movement, and temperature of the water. Vegetation, wildlife, beaches, fishing, and rafting all are influenced by these changes.

How might policy decisions on various flow regimes create risks to all of these resources and how are these risks assessed? As an example, certain flows would allow some warming of water temperatures. The exotic fishery (i.e., trout) is dependent on cool water, while the native fish (e.g., humpback chub) require some warm water for spawning. The two are not totally incompatible but one fishery is at a greater risk when the other is at a lower risk.

Another example deals with high flow releases. Periodic high flow releases may scour exotic vegetation creating a more natural environment. However, scouring with sediment-free water from the dam moves a great deal of sediment downstream destroying beaches which are not replaced. Vegetation loss also reduces bird and wildlife diversity and high flows may create a high risk for river rafting.

The decision to chose a recommended flow, both quantity and timing, must be a compromise. Setting of priorities on resources is difficult because of the many interest groups. It is possible to set flow regimes if enough is known about the interactions of the ecosystem components and processes and response curves can be developed for each resource. Unfortunately, information on some species in the canyon is incomplete. Sediment transport dynamics are also not well known. With these gaps in information, evaluation of risk levels to the ecosystem components is not accurate. Further short- and long-term studies are needed, but meanwhile, risk analyses need to be done for the endpoint resources to facilitate recommendations on flow regimes.

Relationships to EPA Priorities

These two case studies are not closely related to the past priorities of EPA. However, there is an increasing concern for the welfare of the ecosystems on which we are all dependent. These cases demonstrate how, with sufficient information, management or policy decisions can be made based on evaluations of the risks to components of the ecosystem. The Mono Basin study showed a nearly complete assessment while the Glen Canyon study showed an inability to make recommendations because the assessments were incomplete.

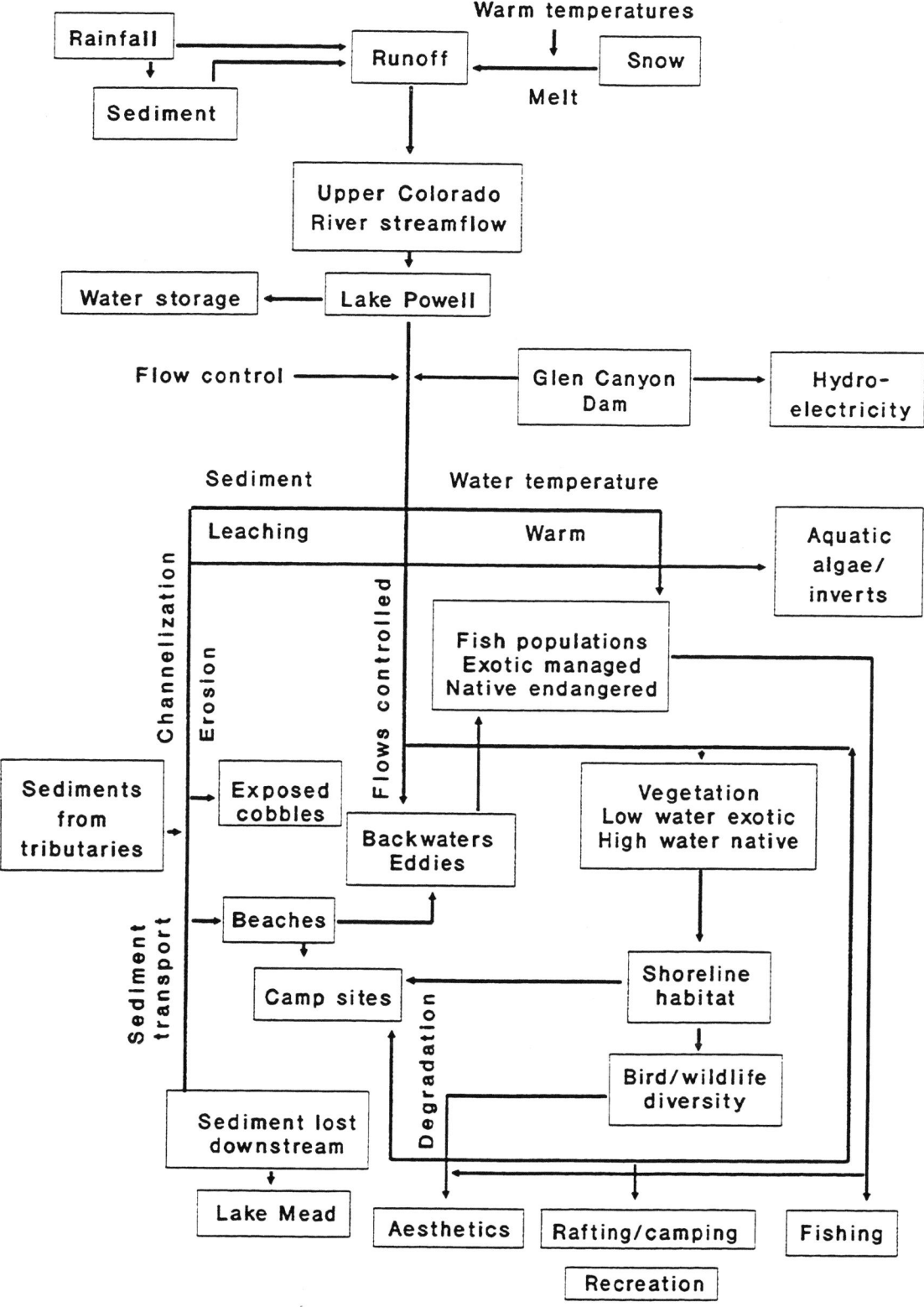

FIGURE 3 Conceptual scheme of Glen Canyon ecosystem components and their interactions under present operations of Glen Canyon Dam. (From National Research Council, 1987. River and dam management, National Academy Press, Washington, D.C.)

Still, recommendations on ecosystems such as Mono Basin and Glen Canyon lie outside the purview of EPA. Let me suggest, that there are circumstances where these systems do fall within EPA's directive. What would happen if underground gasoline tanks at Lee Vining, a town just above Mono Lake, leaked into the groundwater feeding Mono Lake? Is our information base on closed basin lakes like Mono Lake sufficient to project the consequences and take some action? Contamination of bodies of water is always possible.

Should EPA limit itself to contaminants? Is the water quality of Mono Lake or the Colorado River of importance? Dissolved salts and temperature are characteristics of water quality. Variation in these factors that affect the natural biota should be a concern of EPA. Management of our natural ecosystems is controlled by many agencies but the ultimate watchdog of ecosystem health must be the responsibility of one agency and EPA is the only agency whose sole objective is ensuring a healthy environment for humans who are inextricably dependent on natural ecosystems.

SELECTED DEVELOPMENT NEEDS FOR ASSESSING ECOLOGICAL RISK AT THE COMMUNITY AND ECOSYSTEM LEVEL

Terence P. Boyle

Water Resources Division
National Park Service

A background paper for Ecosystem Risk Assessment and Monitoring, March 2-3, 1989, Airlie House, Virginia.

This paper is intended to provide some background information and suggestions for improvements in the art and science of ecological risk assessment and environmental monitoring. Emphasis is focused on research needs in the development of methods for assessments for ecosystem risk using natural ecosystems. It is not intended to be representative of all the needs in this field of endeavor. To date ecosystem risk assessment has been commonly effected by the use of surrogates in the laboratory with batteries of single species toxicity tests, and microcosm tests of various complexities which assess for the effects of chemicals at the population, community and ecosystem level of organization. Microcosm tests performed outdoors containing natural communities or portions thereof have recently come to be known as mesocosm tests and include some of the variability inherent in natural ecosystems. Several generations of computer models have also attempted to assess ecosystem risk by various types of environmental stress mainly considering ecosystem aspects of transport and the fate of chemical contaminants.

There are several levels of consideration in selecting news research needs in ecosystem level risk and monitoring. These fall into three categories: 1) Development of procedures to include the administrative or management aspects of the problem, 2) The use of previous and existing monitoring data and management questions to formulate comprehensive risk assessment and monitoring programs, and 3) Selection or development of technical tools necessary to assess changes at the community or ecosystem level that will answer questions implicit in ecosystem risk assessment and monitoring or ecosystems.

Administrative/management

The first set of considerations that needs to be addressed in collecting environmental data is, "What will be done with ecological data when it is generated?". These are the social, economic, political, and legal aspects of the risk assessment and monitoring that directly affect the scope and detail of any program. For example, the purpose of environmental monitoring can be divided into detecting trends, surveillance, impact assessment, or a legal regulatory aspects. These considerations predicate to a large extent the technical plans for the assessment or program such as what variables are included, which methods are used, the intensity of data collection, statistical resolution, etc. Since ecosystem risk assessment and ecologically based monitoring are predicated on the assumption of anthropogenic activities, the technical information should be planned and formatted for use in the social, political, and legal arena. Scientists tend to design risk assessment schemes and monitoring networks that are technically defensible with other scientists. While technical integrity is an important prerequisite to any scientific program, the development of procedures addressing the social, economic, legal, and political dimension is perhaps the least considered aspect in the design of environmental monitoring programs. Moreover, explicit treatment of management/science problem has received inadequate treatment in the technical literature (Shaeffer et al. 1985 and Perry et al. 1985). This problem is further compounded by the paradox of various agencies with environmental programs at all levels of government seeking to simultaneously cooperate and insure continuity of overall program goals and to differentiate their programs from one another to avoid the appearance of redundancy.

Examples of Existing Programs

The assessment of ecosystem risk has two components: 1) delineation of environmental stress or threats, and 2) determination of ecosystem vulnerability and response. Table 1 lists examples of existing monitoring programs at the global, multinational, national, regional, and local levels. In general, the global and multinational programs address aerial or other transboundary transport and effects of pollutants, national programs address the environmental mandates of their respective agencies, and the regional programs address interagency needs within a recognized physical geographical boundary. If the data from these systems is to be used in any concerted fashion to aid in the formulation or selection of ecologically based risk analysis, there is a need for synthesis of data from a variety of sources. Especially at the regional level, there is usually data available from a multitude of existing monitoring programs that need to be collated and synthesized to understand the potential cumulative impacts on the ecosystems in question. Some of the existing programs listed in Table 1 and elsewhere have areas of geographical overlap and would need to be addressed in some sort of overlay fashion.

Technical

The second component of ecological risk assessment, that is determination of ecological vulnerability and response, has two areas of technical development that are emphasized here. First, there should be some method of determining what ecosystem processes and what types of ecosystems are most vulnerable to outside stress. While this has been done in a relatively simple way for lentic systems exposed to acid deposition by determining the Acid Neutralizing Capacity (ANC), determining what factors contribute to ecosystem 'robustness' needs to be considered in an environment with a multitude of stressors. Some of the properties of ecosystem stability discussed by Westman (1978) may be applicable for ranking the sensitivity to outside stress. Ecosystem characteristics such as total gross productivity, total biomass, redundancy of structural and functional elements, resistance of key species to environmental stress, and resilience as measured by the K or r reproductive strategies of key species, may be the explanation to classifying and ranking ecosystems as to their ability to absorb stress or their ability to recover stress. The EPA's Ecoregions program currently underway is a logical framework for answering these questions (Omernik 1987). Implicit in the Ecoregions approach, especially for aquatic systems (Larson and Hughs 1987), is the designation of 'Reference sites' where the characteristics for 'ideal' aquatic ecosystems within a given ecoregion are determined. I would like to add the concept of a natural resource inventory as elaborated in the chapters of Kim and Knutson (1986) to be selectively conducted within ecoregions considering both the potential stress acting on that ecoregion, as well as the social, economic, political, and legal considerations of the natural resources found there. With additional specifications the properties of ecosystem robustness or sensitivity could also be addressed. Moreover, the problems of temporal and spatial ecosystem variability can be addressed by a resource inventory in the context of the Ecoregions framework. With this resource inventory by ecoregion, a geographical information systems approach could assess cumulative impacts and link land-water ecosystems.

Secondly, the field of ecosystem mensuration is overdue for a review and reevaluation with approaches towards ecosystem risk assessment and quality assurance and control aspects. Ecosystem assessments and responses of ecosystems to stress are most often made at the community level of organization. Community may be defined as the living portion of the ecosystem but in practice communities are operationally defined by the ecologist by the methods of data collection and the statistics that define the data. For example, in aquatic systems there are phytoplankton, periphyton, zooplankton, benthic macroinvertebrate, fish communities etc., each of which is specifically defined by methods of collection, degree of taxonomic resolution, and the statistics used to reduce and interpret the data. Community level indices have been one common method of attempting to reduce the large amount of data inherent in community level studies. Table 2 lists some of the current community level indices in use by ecologists in interpreting community level data. While these indices have been popular among both theoretical and

Table 1. Selected list of monitoring programs of different geographical scope.

I. Global Monitoring System

*International Geosphere/Biosphere Program (IGBP)

*Global Environmental Monitoring Systems (GEMS)

*World Meterological Organization (WMO)

*Man And the Biosphere Reserves (MAB)

II. Multinational Monitoring Systems

*European Monitoring and Evaluation Program (EMER)

*Integrated Background Monitoring Stations (IBMS)

III. National Monitoring Programs

*National Acid Precipation Assessment (NAPAP)

*National Water Quality Assessment Program (NAWQA)

*Environmental Monitoring and Assessment Program (EMAP)

*National Surface Water Survey (NSWS)

*National Contaminant Biomonitoring Program

IV. Regional Monitoring Programs

*Mussel Watch

*Multidecade Monitoring Program of Chesapeake Bay

*Long Term Resource Monitoring Program, Upper Mississippi System (LTRMP)

V. Local

*National Park Service Inventory and Monitoring Program

*National Forest Service

*Other Federal, State, and Local Governmental Agencies

Table 2. Community level indices currently in use in ecological studies.

A. Structural based indices
1. Diversity (8)
2. Similarity (>20)

B. Saprobian based indicies (invertebrates)
1. Chandler's
2. Chutter's
3. Hilsenhoff's

C. Biological based indices
1. Index of Biological Integrity (IBI) (Fish)
2. Insect Community Index (ICI)
3. Biotic Condition Index (BCI) (Insects)

applied ecologists seeking to measure the basic structure analyze the response of natural communities to various influences, the field of community metrics remains relatively poorly technically developed. One fundamental property of community structure is the taxonomic enumeration and how the number of individuals within the community is distributed among them. The structurally based diversity and similarity indices in Table 2 (Washington 1984, Cheetham and Hazel 1969) appear to be differentially sensitive to both the initial structure of the community and also the manner in which the community is changed (Boyle et al. 1989). While the principal deficiency in their application is similar to a Type II statistical error, that is; no response or paradoxical response to rather drastic changes in a community structure. The Saprobin based indices are dependent on the ranking of the relative sensitivity invertebrate taxa to high organic loading or low dissolved oxygen level—sewage impacts (Chandler 1970, Chutter 1972, Hilsenhoff 1977). These indices may give erroneous readings to community change due to environmental stress when the relative sensitivity among the taxa in the community is different than the Saprobian system. These indices also suffer from 'regionalism,' that is the ranking system for taxa sensitivity to the strees is not necessarily easily directly applicable from one region to another. The Index of Biological Integrity (IBI) incorporates a number of biologically interpretable metrics of the fish community into a single index (Karr et al. 1986). This index has been successfully adapted to an number of different regions within the United States. It does appear to be sensitive and able to detect a number of direct and indirect environmental stress. However, one documented potential weakness in using the IBI for assessment or monitoring is that while the fish community is sensitive to a wide variety of environmental conditions the index appears insensitive to the early stages of some type of stress (Berkman et al. 1987). The Insect Condition Index has not received wide enough application or use for evaluation here. The Biotic Condition Index (Winget and Mangum 1979) is an attempt to rank the relative sensitivity of aquatic insect taxa to several environmental parameters in streams. It has similar limitations of applicability as the Saprobian based indices.

BIBLIOGRAPHY

Berkman, H.E., C.F. Rabeni, and T.P. Boyle. 1987. Biomonitors of stream quality in agricultural areas: fish vs. invertebrates. Environ. Management 10:413-419.

Boyle, T.P., G. R. Smillie, J.A. Anderson, and D.R. Beeson. 1989. A sensitivity analysis of nine diversity and seven similarity indices. J. Water Pollution Control Fed. In Press.

Chandler, J.R. 1970. A biological approach to water quality management. Water Pollution Control. 69:415-421.

Cheetham, A.H.J. and J.E. Hazel. 1969. Binary (presence/absence) similarity coefficients. Journal of Paleontology 43:1130-1136.

Chutter, F.M. 1972. 1972. An empirical biotic index of water quality in South African streams and rivers. Water Res 6:19-30.

Hilsenhoff, W.L. 1977. Use of arthropods to evaulte water quality in streams. Technical Bulletin No. 100 U.S. Department of Nature Research 16 pp.

Hughs, R.M. and D.P. Larsen. 1988. Ecoregions: An approach to surface water protection. J. Water Pollution Control Fed 60:486-493.

Johnston, C.A., N.A. Detenbeck, J.P. Bonde, and G.J. Niemi. 1988. Geographical information systems for cumulative impact assessment. Photogrammetric Engineering and Remote Sensing 54:1609-1615.

Karr, J.R., K.D. Fausch, P.L. Angermeir, P.R. Yant, and I.J. Schlosser. 1986. Assessing biological integrity in running waters: A method and its rationale. Special Publication No. 5. Illinios Natural History Survey. Champaign, IL.

Kim, K.C. and L. Knutson. 1986. Foundations for a national biological survey. Association of Systematics Collections 215 pp.

Omernik, j.m. 1987. Ecoregions of the conterminous united states. Ame. Geogr 77:118-125.

Perry, J.A., D.J. Schaeffer, H.W. Kerster, and E.E. Herricks. 1985. The environmental audit ii: Application to stream network design. Environ. Management 9:199-208.

Schaeffer, D.J., H.W. Kerster, J.A. Perry, and D.K. Cox. 1985. The environmental audit i: Concepts. Environ. Management 9:191-198.

Appendix C

Research to Improve Predictions of Long-Term Chemical Toxicity

A Workshop Report

Summary

The development of methods for identifying toxic chemicals has been the focus of substantial research efforts by the National Institute of Environmental Health Sciences (NIEHS), the National Toxicology Program (NTP), the Environmental Protection Agency (EPA), the Agency for Toxic Substances and Disease Registry (ATSDR), the Food and Drug Administration (FDA), and other agencies at various levels of government. This report summarizes a workshop held in Washington, D.C., on December 13-15, 1989, on research to improve the methodology for predicting the long-term toxicity of chemicals.

The salient concepts indicated by a number of workshop participants are:

• Developing improved methods for predicting the long-term toxicity of chemicals will depend on improving our understanding of the underlying science and on effective coordination and integration of the relevant clinical, epidemiological, and laboratory approaches. To achieve these goals and to mobilize needed resources, a national strategy for fostering and coordinating such activities in the public and private sectors is needed. It is encouraging that the National Institute of Environmental Health Sciences (NIEHS) has the beginning of such a strategy in place.

• The existing methods for predicting long-term chemical toxicity—for example, analysis of structure-activity relationships (SAR) and other correlational techniques, in vitro and in vivo short-term tests (STTs), and longer-term animal bioassays—all have valid, if different, uses at present and promise to become increasingly effective in the future, given appropriate research for their further development, validation, and integration. Understanding the mechanisms of action at relevant exposure concentrations and times should improve the accuracy of predictions from laboratory results to humans.

• Computerized SAR analyses—ultimately ideal in terms of their relative economy and rapidity—have considerable potential but have been restricted largely to the prediction of mutagenicity and carcinogenicity of chemicals in structurally related classes. Such analyses can be expected to become more useful within the next 5 to 10 years through improvement of analytical models, better understanding of limitations in the models, lowered computer costs, and widening of the range of toxicological end points to which the models are applicable. These advances will depend to a large extent on: a) further research to validate SAR methodologies; b) expanded research on the use of the methodologies, with particular reference to end points other than mutagenicity and carcinogenicity; c) coordinated development of data bases of the richness and quality needed to support the application of SAR methodologies, including the pertinent metabolic, toxicokinetic, and functional data; d) investigation of the importance of the physical-chemical properties of substances in SAR models; e) studies to explain the molecular mechanisms underlying SAR; and f) provision of stable funding necessary to support the further development of SAR methodologies.

• More than 180 in vitro and in vivo STTs for predicting mutagenicity and carcinogenicity have been developed, but few have been evaluated on a sufficiently broad spectrum of carcinogens and noncarcinogens to define their predictive value, and no single SST or combination of SSTs has been found to be adequately predictive of carcinogenicity. For toxicological end points other than mutagenicity and carcinogenicity, there has been relatively little development of STTs, although for certain life-cycle-dependent processes such as reproduction, STTs might have advantages over long-term tests in the detection of toxic effects. In spite of their currently limited usefulness for predicting long-term chemical toxicity, STTs have increasingly wide use for screening purposes and for supplementing other methods of evaluation. Because they can done quickly and cheaply, such tests promise to become more important in the future. Thus, to expedite STT developments, the following research activities are suggested: a) further validation of STTs for specific uses; b) development on a battery of STTs that are predictive for carcinogens of all chemical classes; c) development of STTs applicable to toxicological end points other than mutagenicity and carcinogenicity; d) development of STT's capable of assessing potency as well as binary (positive or negative) responses; e) further development of STTs applicable to human cells, tissues, and body fluids (e.g., biological markers and DNA probes); f) establishment of an organizational and policy framework to coordinate relevant STT development and validation activities in government, industry, and academia; and g) establishment of training programs and other measures to increase the number of able investigators in the field.

• Of all toxicological methods employing surrogate test systems to predict the long-term toxicity of chemicals for humans, those based on studies using whole animals have the widest applicability and appear to be the most reliable, although the predictability of many animal models remains to be validated. Especially deserving of further development are animal test methodologies for predicting neurotoxicity, reproductive toxicity, immunotoxicity, developmental toxicity, and the toxicological effects of chemical mixtures. To improve these methodologies, the following research activities are suggested: a) studies to elucidate the fundamental biology of toxicological effects on organ systems in different species and strains leading to the selection of the best predicting species and strains, knowledge of which is essential for confident extrapolation of animal data to humans; b) refinement of toxicokinetic and mathematical dose-response models for estimating human risks from animal data; c) investigation of the extent to which toxicological data on a subset of compounds in a given chemical class are predictive for other members of the same class; d) study of the extent to which the predictive reliability of animal models may be enhanced by optimizing the experimental protocols; e.g., through changes in the duration of exposure, age, or genetic background of the experimental animals (including the use of outbred animals or transgenic animals), and e) development of institutional and funding mechanisms to foster integration of studies on toxicological mechanisms into the design and conduct of long-term animal bioassays.

• Because the principal utility of surrogate test systems depends on their ability to predict for the human species, the systematic collection and study of relevant human data are important. For purposes of test validation, therefore, the following research activities are needed: a) clinical, toxicological, and epidemiological investigation of the effects of human exposures to chemicals; b) international cooperation to exploit research opportunities arising out of accidental, occupational, or other types of high-level human exposures wherever they may occur in the world; c) research to improve short term follow-up and hypothesis generation, in human exposure assessment, through government-university-private sector collaboration, computer linkage of poison control centers and occupational health clinics population registries, repositories for storage of appropriate biological specimens, studies of pertinent pharmacokinetics, and evaluation of relevant biological markers; d) improvement in the collection and recording of morbidity and mortality data on the general population; e) systematic study of the pharmacokinetics and toxicological responses of humans, compared with nonhuman species, through carefully designed studies on human volunteers, human tissues, and human cells, including human germ cells; and f) expansion of interdisciplinary training to increase the number of scientists with the kinds and breadth of expertise needed.

Research to Improve Predictions of Long-Term Chemical Toxicity

INTRODUCTION

This report summarizes a workshop held in Washington, D.C., on December 13-15, 1989, on research to improve the methodology for evaluating the long-term toxicity of chemicals. Workshop participants are listed at the end of this document.

The questions placed before the workshop participants were:

1. How can we get information on the toxicity of a material more efficiently, more accurately, and more quickly? This multisided question was addressed by the groups examining structure-activity relationships, short-term in vivo and in vitro tests, and whole animal tests.

2. How can we know that the data developed from nonhuman systems are meaningful for humans? This question was addressed by the group examining methods for validating predictive tests of chemical toxicity.

3. Do the answers to the first two questions most effectively tell regulators and the research workers where next to go scientifically? The group on overarching strategy, policy, and resource considerations attempted to answer this question.

Workshop participants did not attempt an exhaustive scientific report, fully referenced and data rich. Rather, this report is a summary of the joint expert opinions of a collection of specialists in several fields. Not all participants necessarily agree with all aspects of this report.

Several consistent themes appear in the answers provided by the work groups. First, much more work is needed on the validation (in human populations, wherever possible) of the predictions made from the various testing and forecasting systems. For example, many participants in the group examining SSTs found a strong need for determining how well the existing tests perform as predictors. They did not look with any great enthusiasm on devoting equivalent energy to developing new tests. Several participants in each of the working groups expressed a need to develop a unified direction and coordinate research efforts as current programs are scattered. In addition, applied research programs are needed to answer the question, "How could a research program oriented toward a specific publicly obvious goal be moved toward that goal efficiently and effectively?"

Until recently, the major toxic end point examined in the various predictive systems has been cancer. The question asked was usually, "Did this material increase cancer incidence?" Other toxic end points are often harder to measure—but awareness of them exists—and developments are needed so that knowledge of toxic effects other than cancer can enter into a regulator's evaluation of a material. Several workshop participants questioned the adequacy of testing for neurotoxicity, teratogenicity, and related reproductive effects and birth defects, and general organ toxicity. As part of the emphasis on applied science, much more work needs to be done to develop knowledge of the mechanisms or the pathways leading to a toxic response to a xenobiotic.

The questions about the adequacy of testing recall the review *Toxicity Testing: Strategies to Determine Needs and Priorities* (NRC, 1984). Several workshop participants believe there is strong reason for a new look at the problems raised in the report and the recommendations made in it.

The specific considerations and suggestions from workshop participants are given below.

Each group approached the problems in a similar way—defining the current status of the field; considering administrative needs, such as funding and training; and paying substantial attention to unresolved problems, research needs, and potential for further development. Only one group found it useful to prepare an extended bibliography.

STRUCTURE-ACTIVITY RELATIONSHIPS

Status and Outlook

The potential for the study of SAR appears to lie in goal-oriented efforts to develop standardized published data for several toxicological end points in addition to cancer. Several workshop participants indicated that such work could provide the potential for identifying chemicals not known to pose a long-range health hazard. Once the methodologies have matured and have been validated, this could be done efficiently for a large number of chemicals, at reduced expense and without the need to sacrifice a large number of animals, thus making testing goal-oriented and cost-effective. Uses of SAR over the next 5 years call for progressive improvements driven by increased efficiency, lowered computer costs, improved models, and a greater range of biological effects for which structural relationships can be identified. Workshop participants indicated that this approach could lead to greater acceptance, better understanding of the uses and limitations of the models, and increased confidence in their applications, not only to carcinogenicity but also to other areas of toxicology, during the next 5 years.

Research and Resource Needs

Review of SAR Methodologies

Current SAR methodologies stem mainly from similar systematic studies of basic mechanistic research in chemistry. With the availability of computers, these studies have evolved into a promising field of research for evaluating toxicological end points. Although the field is still in its infancy, it provides the possibility of bringing a rational or even a mechanistic basis for understanding the experimentally observed toxic end points.

Problems specific to biological end points still need to be resolved. In contrast with chemical properties, biological end points often are the result of complex components and intertwined mechanistic paths. These complexities include transport properties, metabolic transformations, receptor-binding activity, and reactivity. Models exist for each of the properties and usually enter SAR studies in one form or another. However, these components are considered in parallel, particularly in quantitative SARs. The consideration of hierarchy in the occurrence of biological events is one area in which progress might be made.

Within each of the current SAR models, better understanding of molecular recognition and alternate representation of molecular properties might be considered as high priority. In addition, relevant models and mechanistic relationships than now are available could be identified.

Validation

Several workshop participants indicated that validation of predictions is a major area of concern. Ideally, a validation process should lead into an exploration of the specific deficiencies and defects in current methodologies. Some believe that validation should involve prospective prediction of specific activities (carcinogenicity, teratogenicity, etc.) for a group of chemicals of unknown activity. The process of validation is not a pass or fail examination, but rather part of an ongoing learning process; information derived from a validation effort could be used to improve the next attempts at prediction. Experts might be included in the validation process to attempt to predict long-term toxicity of new chemicals, such as those in EPA's premanufacturing notification process.

Availability of Relevant Toxicological Data

Although much SAR development has concentrated on carcinogenicity (and to a lesser extent on mutagenicity and certain ecotoxicological end points), data are available for teratogenicity, acute toxicity, and other toxicological end points. These data have been only sparsely used; however, a formal mechanism or clearinghouse to collect these data is not needed. Rather, mechanisms could be set up so that individual investigators can be guided toward these sources.

For example, results assembled on genotoxicity led several workshop participants to believe it is possible to develop a battery of SAR models for genotoxicological end points. Inasmuch as chemicals are not usually tested in logical sets, these participants believe it is desirable to use SAR models to identify chemical moieties that need to be tested to fill in gaps in toxicological knowledge.

Treatment of Physicochemical Properties in Models of Long-Term Toxicity

Most SAR models of long-term toxicity, including models of carcinogenicity, have used descriptors of chemical structure to explain the toxicological end points. There also seems to be a place for using physicochemical (p-c) factors, such as octanol-water partition coefficient (log P), solubility, melting and boiling point, dissociation constant (pKa), dipole moment, molar volume, molecular volume, and bond dissociation energy.

Several p-c factors have been used in models of acute toxicity, e.g., log P and bioconcentration factor in models of aquatic toxicity and (pKa) in models of skin and eye irritation. Substantial progress has also been made in the development of pharmacokinetic models. However, the connection between pharmacokinetics and long-term toxicity is lacking for most situations.

In many instances, the use of the substructural fragments that contribute to p-c factors is at least as effective in modeling toxicological end points as the factors themselves; for example, a particular value of log P can be obtained from many diverse structures. On the other hand, it is important to determine whether equally good SAR models could be devised if the p-c factors themselves were used instead of their substructural components.

The Molecular Basis of Toxic Effects

Workshop participants indicated that research should be directed at the molecular mechanisms of action of carcinogenic and genotoxic compounds, as well as for other toxicological end points. This research might include the following:

1. Detailed investigations of the SARs of possible carcinogens and their metabolic intermediates.
2. Synthesis and biological evaluation of key compounds as predicted by SAR methods and mechanistic models to have identifiable toxic properties.
3. Detailed studies of the nature, rates, and mechanisms of critical reactions (e.g., alkylation) of electrophilic species and other intermediates.
4. Study of the role of pharmacokinetics in activation or detoxication.
5. Investigation of the chemical reactivity of "masked" electrophiles or other reactive intermediates that could be unleashed after a biological transformation, such as quinone methides, anion radicals, or cation radicals.

Incorporation of Functional Criteria Relating to Toxicological End Points

Reports or publications in toxicology should include all pertinent data on functional criteria that might aid in evaluation and interpretation of long-term toxic effects and facilitate extension of the results to SAR evaluation of untested compounds. These data include results of short-term predictive tests, transpecies effects, dose/potency factors, chemical disposition, metabolic pathways, acute or subchronic toxicity, and effects on immune or endocrine systems. Such "supplementary" data are not published by many journals due to space restrictions, but several workshop participants believe efforts should be made to provide a microfilm record of such material. Data should be systematically organized on a chemical structural class basis to identify class-specific functional criteria, which would be useful for reinforcing or complementing SAR prediction within a specific chemical class.

Development of Methods for Identifying Nongenotoxic Carcinogens

Better methods are needed to identify what are now called "nongenotoxic" carcinogens. An appreciable number of compounds are considered nongenotoxic carcinogens including most halogenated aliphatics, 2,3,7,8-tetrachlorodibenzodioxin (TCDD), some chlorinated pesticides (DDT and chlordane), 2,2,4-trimethylpentane, di(ethylhexyl)phthalate, benzene, and saccharin. More research is required to systematize knowledge of the factors involved in the ultimate mechanisms of action of nongenotoxic carcinogens and to elucidate the stages in the process. Such efforts will increase the likelihood for better recognition of epigenetic or nongenotoxic carcinogens.

Funding

Much SAR development stems from studies conducted in universities. Workshop participants believe new approaches must be evaluated and encouraged and funding developed to bring new talent into this area. Research proposals submitted to granting agencies often are too narrowly focused to be funded. As a result, very few groups have been involved in such research. Collaborative or cooperative efforts between government and academic institutions should be encouraged to promote more efficient use of resources.

SHORT-TERM IN-VIVO AND IN-VITRO TESTS

The major effort to date in the development of short-term tests (STTs) has been in the area of genetic toxicity testing for the prediction of carcinogenicity after observation of mutagenicity. More than 180 short-term in vitro and in vivo tests have been developed or proposed to detect rodent carcinogens and, by implication, chemicals that would be carcinogenic in humans. Test combinations (or batteries) and sequential testing schemes have been proposed as ways to improve the predictability. In general, the batteries and combinations have not proved to be better predictors than the best of the single tests. Too often batteries of tests have increased sensitivity (finding more positives) at the cost of decreased specificity (labelling materials positive that were really negative).

STTs have been developed for other toxicological end points; however, the level of investigation for the majority of these has not been as extensive as that for carcinogenesis. Therefore, STTs for carcinogenesis can be used as the paradigm for all SSTs in general, and the mistakes made and lessons learned in this field can be used to guide development of comparable tests for other pathological end points.

Properties, Uses, and Limitations of STTs

STTs often are defined with respect to the long-term effect they are being used to identify or measure, rather than to a specific time component. For the purposes of this report, a precise definition of a STT is considered to be less important than the general principles that workshop participants believe could guide STT development, validation, and use. There are four reasons to perform an STT:

1. For screening purposes, to replace more costly long-term testing or to provide information that could be used to help design subsequent tests.
2. As an adjunct to other tests to aid in making regulatory or industrial decisions or to aid in the elucidation of mechanisms of toxicity.
3. To replace existing tests with more facile tests.
4. To take advantage of windows of opportunity for special toxicological responses involving life-cycle phenomena (e.g., embryonic development and fetal maturation) that might be missed by traditional testing in a standard population.

There are two types of STTs—those that are surrogates for other tests and measure the same toxicological event as the tests they replace and those that can be correlated with other toxicological end points, assuming little or no knowledge of mechanistic relationships. As with most biological systems, some STTs will not fall neatly into one or the other category but will have features of both.

STTs need to be validated. Like any other predictive tests, STTs will give false positives and false negatives. If STTs are part of a public-health-oriented scheme, attempts to develop a "no false-negative" screen might be justified. Such a screen or test would have to be followed by more sophisticated tests capable of identifying false positives. Similarly, a materials development program (such as the discovery of an anticancer drug) might require that the test uncover all potentially useful materials with subsequent tests designed to weed out false leads.

Validation leads to the determination of the ability of a test to fulfill its intended purpose. The usefulness of a test often is constrained by its inter- and intralaboratory reproducibility of the test and the performance of the test under controlled conditions using coded chemicals. In addition, workshop participants believe that the effectiveness of a test for discriminating between toxic and nontoxic chemicals could be determined, and the results of the validation studies published in the scientific literature. One important finding of the validation process might be that a test were accurate for some chemical classes and not for others.

Research and Resource Needs

Recruitment and Training of Personnel

The development of any technique, no matter how simple or straightforward, requires trained personnel. STTs usually are categorized as "applied research" and often are not seen as attractive by academic researchers who are the mentors of young research workers. However, for public-health concerns, research into the use of STTs might be as vital as basic research. Workshop participants believe programs should be developed and funds provided to interest researchers in STTs and prediction of long-term toxicity and to provide training in these fields.

Application to Nonbinary Toxicological End Points

STTs and long-term tests have been used to classify materials on a binary basis—yes/no, carcinogen/not a carcinogen. This has made comparisons among tests straightforward and relatively easy. However, it is obvious that many toxicological end points should not be expressed in a binary fashion, but rather as dose-response curves, dose ranges, or on a continuous response scale. For continuous responses, the determinations of sensitivity and specificity are similarly more complicated but are clearly possible. The continuous scales require scientific judgment to establish cutoff points below which a response is considered negative and above which is considered positive.

Coordination of Test Development and Validation

The development of predictive tests has usually lacked coordination. Workshop participants indicated that there is a need to devise an administrative framework to coordinate or direct the development and validation of new STTs. Among the tasks suitable for coordination are evaluating the tests (in use or proposed for use), assembling a centralized information data system that would be continually evaluated and updated and made accessible to investigators who are interested in test development and validation, and identifying areas of toxicology and public health that could benefit from the use of STTs. A coordinating group could identify model chemicals for study as well as develop STT data bases.

Assessment of Applicability to Regulatory Uses

One major use of STTs is to provide data in support of regulatory decisions. The use of STT data by different regulatory agencies should be examined, and the effect the tests have on regulatory decisions should be evaluated. The various regulatory agencies need to codify for public review their current requirements for accepting STTs and the minimum requirements they have for a new STT before the data are acceptable for regulatory purposes.

Application to Human Specimens

Because the ultimate organism of concern to humans is the human, the most relevant STTs would be those that could be performed using biological samples from exposed populations or populations otherwise at risk for chemically induced disease. Several workshop participants believe that STTs to identify and quantify human risk should be developed. These tests could be used as markers of chemical exposure in humans or to identify early disease or organ damage. In addition, investigators developing or using STTs should examine the scientific basis for using cultured human cells in lieu of rodent cells or microbes. Special attention should be given to the identification and implementation of techniques derived from the new developments in biological technology that can be applied in humans and laboratory animal models (thereby reducing the uncertainty in the extrapolation to humans).

Broadening the Spectrum of Toxicological End Points

At present, many STTs are developed around one highly specific end point associated with a specific molecular or physiologic mechanism. Workshop participants believe that the development of short-term biological predictive systems that can be integrated with other systems, measure more than one end point, or respond to more than one mechanism of action should be encouraged.

Periodic Re-evaluation and Revalidation

For the purposes of evaluation, STTs are related to a "gold standard," that is, a specific toxicological effect of chemicals measured in rodents or humans. Workshop participants indicated that this standard, whether it is carcinogenesis in rodents or diminished fertility in humans, may be a poor one. Therefore, all STTs must be re-evaluated from time to time.

More Adequate Reporting of Relevant Data

To aid in the evaluation of STT performance, workshop participants suggested that individuals and laboratories reporting results should also provide quantitative data or make them available on request (including such estimated measures as the Lowest Effective Dose or Highest Ineffective Dose), Chemical Abstracts Services Registry Numbers, structures of test chemicals, and the source and purity of the chemicals tested.

Extrapolation to Humans

Research is needed on methodologies to extrapolate from in vitro to in vivo effects in laboratory animals and in humans without involving the unnecessary exposure of humans to potentially toxic materials.

Further Problems and Uses of STTs

Currently available tests can detect most genotoxic carcinogens. However, increasing evidence shows several types and classes of carcinogens are not detected using existing in vitro genotoxicity tests. STTs are needed that will effectively identify these carcinogens.

It is difficult to evaluate the toxicity of complex mixtures. In vivo, long-term experimental approaches to studying mixtures are severely limited by the high cost of multifactorial experimental designs. STTs have a significant advantage in investigating chemical interactions and the toxicity of complex mixtures. Research should be focused on developing and using STTs to identify chemical interactions and elucidate toxicological mechanisms involved.

More Effective Coordination of Test Development and Validation

No formal science policy or framework organizes and coordinates the use and development of STTs. At present, methodology development is mainly driven by the goals of individual research laboratories. For a program to be efficient in the transfer of new technological developments from the research laboratory to routine toxicity testing, a framework capable of fostering this process is needed. Such a framework has been described by Drs. J. Frazier and A. Goldberg of the Johns Hopkins Center for Alternatives to Animal Testing and is briefly described below.

The overall scheme is divided into two major functional units (Figure 1). Above the dashed line in the figure, the structure describes a framework for test method development and validation. The components below the dashed line relate to judgments of whether a given methodology is acceptable for the purpose for which it is proposed and to place a "seal of approval" on the tests. At the core of the methods development activity (above the dashed line) is a scientific advisory board (or boards). This board would:

1. Review the state of SSTs, identify gaps in scientific knowledge, and recommend research.
2. Recommend testing methodologies to a scientific review panel for evaluation and approval.
3. Advise and guide a chemical bank; select representative chemicals for test method validation.
4. Oversee a toxicity testing data bank and review data prior to its entry into a data bank.
5. Recommend activities for reference laboratories, including validation of developed methodologies and development of orphan methodologies.

The chemical bank is designed to address problems that retard test development and would be a repository for chemicals that could be used by research laboratories and reference laboratories to validate methods. Chemicals would be designated for different methodologies related to particular toxicities or pathologies (e.g., neurotoxicity, hepatotoxicity, and carcinogenicity) for validation and for calibration.

A toxicity testing data bank could collect and catalog data for methodology validation. Data on the performance of proposed and in-use test methods would be collected and classified. If a method were under consideration for validation, the data matrix could be searched for experimental data from all chemicals tested by the particular method. A scientific advisory panel would determine, on the basis of the available data, whether the method could be recommended for consideration by the scientific review panel.

Reference laboratories would serve two major functions. The first function relates to determining how readily a methodology could be transferred among laboratories. In the proposed framework, reference laboratories, under the direction of a scientific advisory panel, would implement the method under consideration and test sets of chemicals from a chemical bank. When sufficient data are collected, the method would be recommended to a scientific review panel for evaluation. The second function of reference laboratories would be to develop orphan methodologies—scientifically valuable tests identified by a scientific advisory panel which have not been fully developed for various reasons. Reference laboratories could be assigned to develop these methodologies further.

The actual evaluation and recommendation of methodologies for specific testing purposes would be the responsibility of a scientific review panel. The panel could be either an interagency panel established by concerned regulatory agencies to evaluate methodologies for general use or individual panels established by each agency to evaluate methodologies in the context of a specific agency's needs. Implementation of such panels and their functions would require cooperation among all interested groups government, industry, and academe.

WHOLE-ANIMAL TESTS

In safety assessment and hazard evaluation of xenobiotics, what are the prospects of improving predictors of toxicity and their translation into risk? What are the relevant research needs and research opportunities to enhance the existing processes, and what are the strategy, resource, and policy considerations involved in enhancing predictability?

Workshop participants unanimously agreed that animal studies are the foundation of health risk assessment and will remain so for the foreseeable future. Assays have served well in protecting public health, but they can be improved. The results of such studies are used throughout the world to regulate xenobiotics in food, air, and water and to assess the safety of ingredients in pharmaceuticals, cosmetics, and pesticides.

The Strategies of Whole-Animal Tests

Historically, as studies in animals have evolved, procedures have become increasingly sensitive and specific. The intact animal has provided the only source of data offering acceptable predictability as the integrator of dose, response, repair, and recovery, including: 1) biotransformation, pharmacodynamics, and pharmacokinetics; 2) compensatory and amplifying processes; 3) aging, including fetal life and geriatrics; 4) cumulative

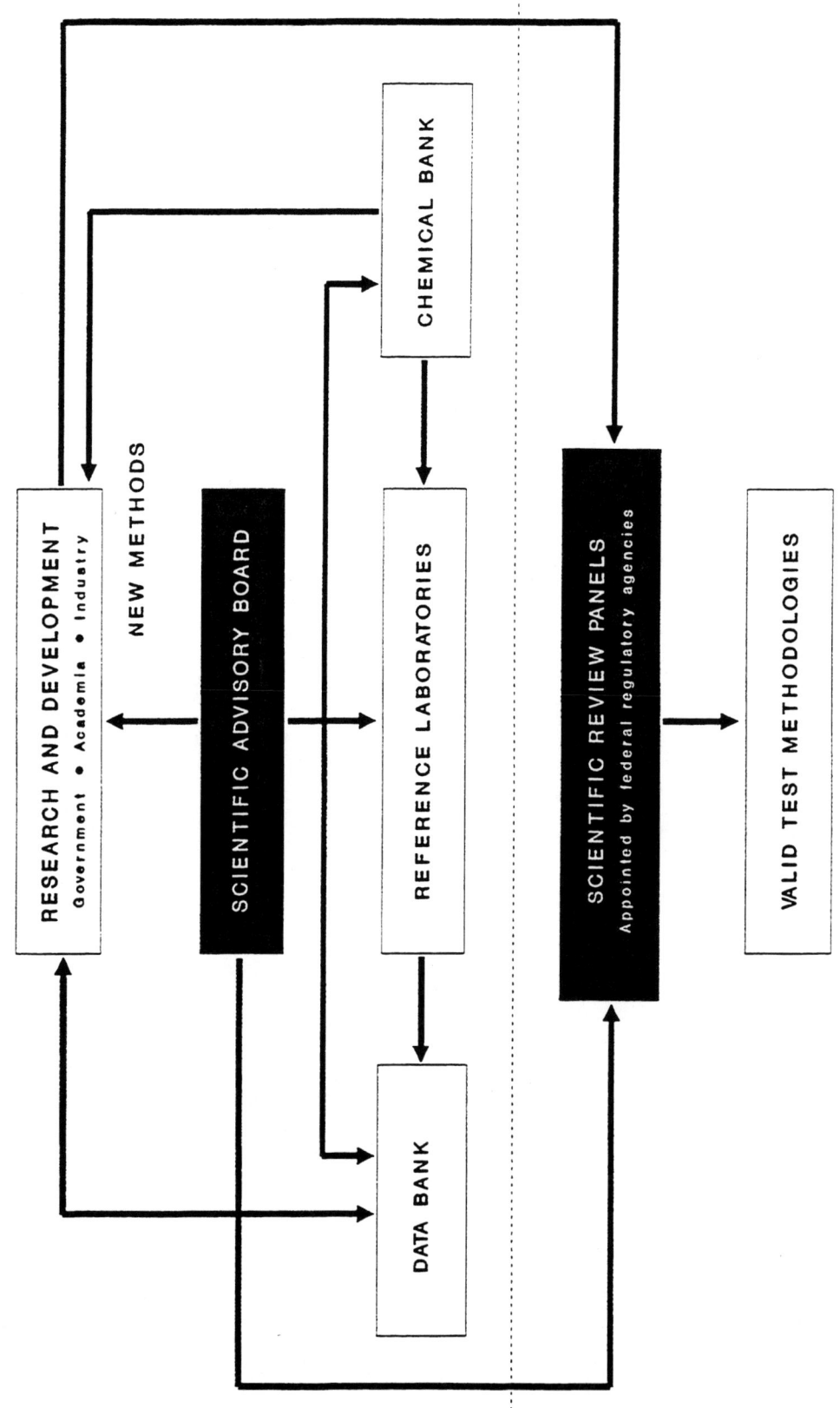

FIGURE 1 Framework for organization and coordination of STT development.

toxicity at the molecular, cellular, tissue, and organ levels; 5) route as a determinant of toxicity; and 6) repair and reversibility.

Workshop participants ranked several toxicological end points to assess relative confidence in the three test systems for predicting adverse outcomes in humans. The systems compared were animal bioassays, in vitro assays, and SAR analyses. For all end points, intact animal studies coupled with available human data were considered to surpass the degree of predictive confidence generated by SAR and in vitro assays. The only area participants believed that SAR and in vitro assays resulted in greater than a 25% reduction in uncertainty was in determining potential carcinogenicity. In this case, the SAR and in vitro observations were judged to give consistent results within individual classes of chemicals (Table 1).

RESEARCH AND RESOURCE NEEDS

Validation by Appropriate Studies in Human Subjects

The art and essence of predicting teratological effects in humans from results of laboratory investigations have improved in recent years. However, in view of the recent activity in this field and conflicting statements, workshop participants were generally unwilling to make firm confidence statements.

Human data for validation are very hard to come by. Some workshop participants indicated that a strategy should be developed under strict ethical guidelines to study selected chemicals, under very stringent conditions, in single administration protocols, in human volunteers. The guidelines should be similar in intent to, but

TABLE 1 Confidence in the Test for Predicting Effects in Humans

Outcome	Structure-Activity Relationships	Short-Term Tests	Animal Tests
Biotransformation PK/PD	3*	2	1
Compensatory Processes	3	2	1
Aging	3	2	1
Cumulative Toxicity	3	2	1
Route/Toxicity	3	2	1
Repair/Reversibility	3	2	1
Teratology	3	3	1
Cancer	2	2	1-2

* 1, highest; 3, lowest.

more rigid in control than, FDA guidelines for the protocols of Phase I experimental trials. Such observations of humans could provide essential information on the qualitative and quantitative aspects of human disposition of a xenobiotic and its metabolites. Human observations also allow laboratory animal models with the closest correspondence to human metabolism to be assigned to further investigations. Accidental exposures also provide an opportunity to gather applicable human data. With the decrease in emergency response time, several workshop participants believe that the scientific community should be prepared to collect samples from emergency incidents.

Funding

According to some workshop participants, research funding related to toxic chemicals is sometimes fragmented or duplicative. Furthermore, public monies are inappropriately spent on toxicological research for proprietary chemicals that are actively marketed. In addition, major flaws in funding priorities and mechanisms impede the application of integrative approaches involving all relevant disciplines. Current funding mechanisms actively encourage fragmentation of research approaches and tend to make toxicology a cottage industry.
Parkinsonism in young drug users. These and other instances of neurotoxicity provide an opportunity to examine questions typically not addressed in the conventional risk assessment scheme, e.g., 1) patterns of toxicity including the spectrum and sequence; 2) rate of progression and reversibility; and 3) latency.

Broadening of Risk Assessment End Points

According to several workshop participants, the risk assessment models for end points other than cancer, such as neurotoxicity, teratogenicity, and general organ toxicity have not been given adequate attention. Functional teratology provides an example of a need for a quantitative decrement model: environmental toxicants such as lead and methyl mercury have already been identified. Another example of chemically induced neurotoxicity is the case of MPTP, the "designer" drug that was found to induce rapid onset Parkinsonism in young drug users. These and other instances of neurotoxicity provide an opportunity to examine questions typically not addressed in the conventional risk assessment scheme, e.g., 1) patterns of toxicity including the spectrum and sequence; 2) rate of progression and reversibility; and 3) latency.

Re-evaluation of Risk Assessment Models

Participants indicated that assumptions underlying risk assessment methods and models need periodic re-examination. Recent advances in toxicology and supporting disciplines show that the nonthreshold linearized multistage model for carcinogen risk assessment might need to be replaced by a more mechanism-oriented model for some carcinogens. The NAS/NRC Committee on Risk Assessment Methodology is making such a reassessment. Similar critical attention to the models used for risk assessment for other end points is needed (e.g., reproduction and development.)

Elucidation of Causative Mechanisms of Toxicity

The development of knowledge in toxicology has been an interactive process, usually commencing with an observation of an adverse effect in humans or laboratory animals, followed by in vitro or in vivo experiments to define the sites and mechanisms of action. Eventually, long-term studies are conducted in intact animals. Predictive toxicology evolved as a modification of this process. The underlying assumption of predictive toxicology is that animal studies, properly designed, can predict biological effects in humans. To improve long-term bioassays, several workshop participants indicated that greater understanding of the mechanisms of toxic effects is needed. Research to elucidate such mechanisms is critical to build confidence for the use of animal test results for human risk assessment. Research on mechanisms should exploit the latest concepts from the entire literature, not just the toxicology-related literature. For this purpose, a series of staggered-start, overlapped investigations might allow an investigative team to describe lesions at the minimal toxic dose and to address the question of mechanisms in the whole animal.

Grouping of Chemicals by Mechanisms of Toxicity

The burden of testing might be reduced if the toxic effects of a group of xenobiotics could be explained by a common mechanism. This approach could limit the number of future tests needed. Further efforts should be made to identify categories or classes of chemicals that induce similar lesions. Representatives of these categories could then be tested in long-term assays to search for further commonalities; that might then further reduce testing.

Participants recommended testing categories for two reasons: First, the NAS Committee on Toxicity Testing Needs noted that fewer than 20% of chemicals in commerce have been tested adequately for toxicity. For some toxicities, life-time testing is not needed and involves unnecessary exposure to some compounds. Second, bioassays are expensive. If results from testing one or a few compound could be applied to an entire class, some workshop participants believe this would enhance the cost-effectiveness of testing and might provide a stronger basis for developing SAR prediction systems.

Methodologies for Complex Mixtures

There is an acute need for methodologies to study complex mixtures (NRC, 1988), including diets that contain naturally occurring carcinogens and anticarcinogens. Studies of food extracts are likely to lead to inappropriate conclusions regarding the toxic potential of fruits and vegetables. Complex mixtures are best studied within the milieu in which they are found (i.e., "top down") rather than as a summation of individual components (i.e., "bottom up").

Studies in Animals with Special Phenotypes and Genotypes

To improve and widen the predictivity of animal assays, some participants called for studies in animals with specific phenotypes and genotypes. Examples include the aged animal (to coincide with the effect of an aging population); for developing and validating biological markers of aging, apart from chemically induced degenerative lesions (defined as structural and/or functional decrements); outbred strains for toxicity testing to mimic the heterogeneous human population; and transgenic mice with human genes, to study susceptibility of the human genes in situ to environmental toxicants, with emphasis on the role(s) of oncogenes, tumor-suppressor genes, and cellular receptors (such as the epidermal growth factor receptor and the hormone receptors) in the toxicity spectrum.

Integrated Analysis of Dose-Response Relationships

Some workshop participants believe that the design and interpretation of whole-organism dose-response studies should be coordinated with dose-response studies of cellular and molecular processes. Models involving possible precursors, such as cell proliferation, need to be further developed.

VALIDATING PREDICTIVE TESTS

Status and Outlook

The major question asked of laboratory tests is how well they predict what would happen to exposed humans. A good predictor is said to be valid test; a valid test will produce few false negatives and few false positives. Toxicological tests and procedures can be used to predict the risk of illness, e.g., the structure of biochloromethyl ether led to the prediction from structure-activity considerations that it was a likely carcinogen; tests in animal systems supported this conclusion, and cancer was found in exposed workers. Ideally, the results of a test should be used to prevent illness. However, when test results preclude the use of chemicals, there will be no human exposure, and true validation, particularly the identification of false positives, becomes impossible. When positive, a test should be used to eliminate or reduce exposure, and thus help to prevent illness. When test results are used to eliminate all exposure, there will be no human data for comparison with the test data, and, as a consequence no opportunity for validation or for identifying false positives. False negatives can still be uncovered, but no formal system is designed to do so.

Technical attempts to reduce experimental variability of all tests used, in the sense of basic quality assurance, is essential. How reproducible is the test? What are the intra- and inter-laboratory variations? What variation occurs over time? Validation of individual tests must be carried out before comparison between tests can be attempted. Validation can be quantitative or qualitative. The process is iterative. On the qualitative level, a test, for example, might be able to identify carcinogens but be unable to quantify their dose-response relationships. Three levels of predictive testing have been identified in this report, ranging from simple to complex situations: level 1) chemical action: SARs; level 2) in vitro and short-term in vivo assays; and level 3) animal bioassays. The results of these may be compared to data from human epidemiology.

Several workshop participants indicated that validation needs to be carried out within and between levels of testing. Many in vitro and short-term in vivo tests have been developed. One approach to improving prediction is to use a battery of tests and, by way of validation, to compare the results of the battery with some accepted standard—recognizing that at times the standard itself may be inadequate. In one study by NIEHS, 73 chemicals were run through four tests (Ames, sister chromatic exchange (SCE), chromosome aberration, and mouse lymphoma) and compared with the outcomes of rodent bioassays. The results did not support the intuitively appealing idea that a battery of tests would improve predictability. The battery was no more valid than the single Ames test.

Validation of the predictability between levels of testing, especially for predicting results between species, presents the most difficult problems. Some laboratory results have, as yet, no meaning with respect to human outcomes. What are the consequences to human health of SCE, or somatic- and germ-cell mutation? Studies in which individuals (human and other animals) are followed for a lifetime are rarely carried out, e.g., following animals with known chromosome breaks to see if more tumors develop in these individuals than in those that are free of the defect.

Knowledge of mechanisms could help bridge the gap between testing levels. Lack of knowledge of exposure is often a stumbling block in attempting comparison of laboratory results with outcomes in humans. In tests carried out at levels 2 and 3, the exposure is known, but predictions between level 3 (animal bioassays) and human epidemiology often founder because human exposure is usually not quantified. Even with knowledge of the slope and shape of dose-response curve, knowledge of exposure is required before predictions can be made. Exposure may at times need to be estimated from knowledge of levels in food, air, and water from direct chemical measurement or from use of biological markers.

In some circumstances, validation of the results of animal experiments by comparison with effects on humans may not be possible given current experimental procedures. According to workshop participants, animal feeding studies are usually begun at a later relative age in the animal than the age at which exposure begins in the human. Exposures to the fetus are not mimicked in laboratory studies except rare multigenerational studies. Additional difficulties arise when the primary exposed individual is the sire and the material under suspicion is delivered indirectly, as a one-time event, to the egg cell (and thus to the embryo) during fertilization. Difficulties in validation between testing levels and the need to have accurate exposure information—which varies greatly with demographic factors such as age, occupation, and location-caused a number of workshop participants to suggest that a major effort should be made to improve epidemiology, particularly of accidentally or occupationally exposed persons. The genetic heterogeneity of the human species in comparison with the highly inbred experimental systems must always be keep in mind. The ultimate link in the validation of SAR, STTs, and animal bioassays is the relationship of the test results to the action of chemicals in humans. The greater the range of materials that have been well studied in humans, the greater the opportunities to validate (or discard as invalid) toxicological tests.

Research Needs and Possibilities

Studies of Heavily Exposed Populations

In an effort to expedite validation studies through comparison with effects in humans, several workshop participants indicated that efforts are needed to make the most of research opportunities arising from massive exposures to chemicals. Any special exposures to humans need to be brought to researchers' attention. Massive ingestion of a pure chemical, for example, for therapeutic reasons, by personal or occupational accident, or by suicidal intention, might provide comparative toxicologists a basis for developing information on biotransformation, tissue distribution, and pharmacokinetics that might guide interpretation of studies conducted in experimental systems. Interaction with clinical toxicologists, poison control centers, state health departments, the Agency for Toxic Substances and Disease Registry (ATSDR), and the National Toxicology Programs (NTP) would expedite examination of valuable research specimens and could improve medical management. For example, studies of reproduction in cancer patients who received well-documented, high doses of genotoxic drugs and radiation provide an opportunity to understand human germ-cell mutagenesis, while validating other measures of somatic cell mutation.

International Epidemiological Studies

Several workshop participants indicated that U.S. support should be encouraged for epidemiological studies wherever there are populations that receive substantial environmental or occupational exposures to chemicals of public-health interest. International studies should be done in collaboration with WHO and appropriate groups in this country (e.g., DHHS, EPA, and NAS-NRC). Exposures should be well characterized and of a magnitude at which observable effects would be expected. The use of biological markers for exposure, biological effects, and susceptibility should be encouraged. Living conditions in non-U.S. study populations should not be so different from those in the United States as to constitute a problem in extrapolating the data from country to country.

Registries of High-Level Exposures

Development of registries of individuals with high exposures, much in the manner of ATSDR, should be encouraged according to a number of workshop participants. Populations with large exposures—due to industrial accidents, for example—might provide the best groups to be followed. Most surveillance systems (e.g., epidemiologic studies, and disease registries) for human disease end points lack appropriate information on exposure. Registries of effects are most useful when they can be linked with exposure information. A problem inherent in following exposed populations for ultimate health effects is that industrial populations are often exposed to many potential toxicants, making the establishment of cancer-causing linkages to single agents difficult.

Improvement of Exposure Measurement and Quantification

To improve predictions of long-term chemical toxicity, priority must be given to developing effective programs in exposure assessment. These could include the following:

1. A computer-linked network of poison-control centers or occupational health clinics would identify individuals who have had biologically significant exposures to environmental or workplace chemicals. Linking this system to a sample storage program that contained samples of body fluids or tissues from selected individuals would enable researchers to approach recently exposed (24-48 hrs) individuals or to obtain samples from individuals with less recent (e.g., up to 1 year prior) exposures. Workshop participants indicated that specific protocols should be developed to be used in the cases of chemical accidents. Such a network/sample storage program could be used to develop information in humans on: pharmacokinetics (metabolism, deposition, and tissue distribution) of an agent; biological markers of exposure; and biological markers of effect.

2. A collaborative program among the government, universities, and the private sector could be used to develop accurate exposure

records on high-exposure employees in targeted segments of selected industries.

Biological Markers of Exposure and Effect

Clinical measures of chemical exposures in humans (e.g., biological markers) need to be related to disease outcome.

Laboratory documentation of exposure to environmental agents has usually been correlated indirectly to suspected health outcomes. For example, various chromosomal defects have been seen as constitutional, and acquired abnormalities in peripheral lymphocytes and bone marrow cells following some chemical or radiation exposures. Some chromosome breakage states have been associated with an excess of malignant neoplasms. But, even among the Japanese survivors of the atomic bombs, direct documentation has not been made that individuals with excessive chromosomal defects are the individuals who later develop excess cancers. With international collaboration, especially where record linkage is excellent, it seems feasible to follow up occupational cohorts that have been karyotyped to document the actual disease outcomes in individuals.

The best predictor of a received dose at present is measurement of the chemicals in human fluids and tissues, e.g., urine or blood. Measurements of chemicals in the environment (e.g., soil and water) are poor indexes of doses to humans. Based on tissue-response information, it might be possible to develop better models to assess exposure by indirect means for chemicals that cannot be measured in human samples either because analytical methods do not exist or because the chemicals are rapidly metabolized and excreted, leaving no traces.

Tissue data could lead to discovering which environmental pathways are important for human exposures. Characteristics of the chemicals (solubility, chemical form for metals, and reactivity), binding to media (e.g., soil, sediment, and particles), particle size, and environmental fate and transport must be considered.

Many chemicals are ubiquitous in our environment in low concentrations, and the general population is exposed to them. Unless subpopulations have received much higher doses, exposure registries are not likely to be useful for the study of disease outcomes. The usefulness of a registry will depend in part on the potency of the chemical, the range of doses received by different individuals, and the heterogeneity of the population. Therefore, several workshop participants suggested that criteria related to these concerns be developed and applied to the establishment of any future registries and to re-examine the usefulness of existing registries.

Molecular Approaches

Workshop participants believe molecular techniques must be developed for extension to epidemiological studies. As new molecular techniques become available for assessing health effects, they should be implemented in epidemiological studies. Two examples are given below.

1. Assays for genetic changes—Better methods are needed in humans for rapid analysis of genetic changes, including recombination, structural, and numerical chromosomal aberrations, as well as genetic mutations. New candidate methods include the use of antikinetochore antibody in the micronucleus assay and mutational spectra analysis. Efforts to implement the developing information on oncogenes and tumor suppressor genes for new epidemiologic assays should be encouraged.

2. Mechanistic markers for reproductive toxicity in humans—Molecular markers are needed for each of the successive stages of the reproductive process. For example, a marker for successful implantation of the fertilized egg (level of human chorionic gonadotrophin, hCG) permits the examination of the prevalence of postimplantation loss. An analogous measure is needed to assay for preimplantation loss. An assay for levels of the hormone relaxin is a candidate approach. The relationship between abnormal sperm factors (e.g., morphology and motility) and clinical outcomes needs to be determined. Linking sperm factors with reproductive risk and developing methods to detect preimplantation loss would enable to characterization of the major determinants of reproductive failure in humans.

As new methods are developed and validated, resources should be made available to permit the full use of these assays in epidemiological studies.

To illustrate the scarcity of resources, all research groups using the radioimmunoassay for hCG to monitor for reproductive outcomes in humans are dependent upon one laboratory for a supply of antibody and for continued development of the technique.

Information on Disease Incidence

Baseline information on the background frequencies of health effects or their molecular surrogates in human populations needs to be developed. Information is usually needed on the effects of age, diet, and other factors on the measured factor. The background frequencies of many clinical end points (e.g., spontaneous abortions and menstrual disorders) also need to be developed.

To improve our understanding of human disease, it is important to know the incidence of specific diseases in the general population and the trends in disease incidence. At present, U.S. record systems are not adequate to supply accurate information. This is particularly true for the less-common diseases, for diseases not usually fatal, for neurological diseases and for reproductive outcomes. Many birth defects are not recognized at birth and are therefore poorly recorded. The number of autopsies conducted has greatly decreased over the past few decades, suggesting that information on causes of death might be becoming less accurate.

Training

Progress in improving predictions of long-term chemical toxicity, as well as related environmental and toxicological issues, will be seriously slowed according to some workshop participants as long as there remains a dearth of scientists with relevant interdisciplinary training. To illustrate, a review of the cytogenetic literature suggests that cytogeneticists often lack epidemiological training. The human studies they plan and conduct are at times undertaken without adequate attention to the sample size, the appropriateness of controls, or the potential for selection biases. Epidemiologists, on the other hand, often lack familiarity with genetic toxicology literature.

Suspected mechanisms often are not known to epidemiologists, and the overall synthesis process is negatively affected. The need for incorporating current developments in oncogene research into epidemiology has already been noted. It would be appropriate for federal agencies to undertake traineeship or postdoctoral programs along the NIH model to develop the needed researchers with interdisciplinary skills.

STRATEGY, POLICY, AND RESOURCES

Linking and Integrating Evaluative Approaches

All of the different toxicological approaches discussed in the workshop have value, and all have reached what might be considered adolescence, if not full maturity. The challenge now is to link them, evaluate their potential in predicting human-health effects, and integrate their selection and use in evaluating particular chemicals (see e.g., NRC, 1983; Anderson and Ehrlich, 1985; U.S. Department of Health and Human Services, 1986).

Several workshop participants indicated that this linking, evaluation, and integration needs to be conducted in perspective of the ultimate uses to be made of the information, for it is obvious that toxicological testing approaches and regimes should be oriented to providing data that will satisfy clearly conceived objectives of human-health protection.

Procedure flexibility, interpretive judgment, and continual improvement of methods are essential. Regulators and other users of tests must be open to adopting improved methods and to applying judgment in making inferences about human-health effects from routine tests. The challenge is to embrace flexibility and methodological evolution within the standardized procedures that bureaucratic systems naturally require (NRC, 1984).

Two themes pervaded the workshop. First, wherever possible, methods should be extended to cover a wide variety of health-effect end points, not only acute injury, cancer, neurological, and reproductive effects, but also immunological, renal, hepatic, and other health effects. For each end point, this implies major research needs. Every effort should be made within existing structures to broaden the reach of tests, rather than trying to develop entirely new administrative structures for the purpose.

Second, several workshop participants indicated that the ability of all of the methods to predict human health effects needs to be validated. Again, to do this properly implies major programmatic needs.

An important way to facilitate linkage and comparative evaluation might be through improving toxicological data bases and data-base systems. Data gaps have to be bridged, and throughout, data quality and relevance have to be established. In all of this activity, the power of the inferential logic that these methods provide is at issue (Hattis and Kennedy, 1986).

Structure-Activity Relationships

Clearly, SAR analyses are informative. But because they have been applied extensively to only a few compound classes, they lack confirmed predictive power for many classes of compounds. Also, they have been much more highly developed for carcinogenicity and a few acute health effects than for other health effects (Ashby et al., 1989).

Several workshop participants indicated that SAR techniques should be extended to many other molecular groups and health-end-point classes. This will require much more data gathering, evaluation in the SAR framework, and validation of predictive power against data from other test systems and biological understanding.

SAR development would benefit from more active coordination among the relatively few SAR research groups and more orientation to the needs of regulatory and other end users of the techniques. Some workshop participants indicated that SAR practitioners should explore coordinative possibilities with the NRC, federal agencies, and other central organizations.

Short-Term In-Vitro and In-Vivo Tests

As with SAR techniques, STTs have been successfully developed for mutagenicity (which can be taken as possible indication of carcinogenicity) and for some acute toxicities, notably neurotoxicity. But, in general, STTs vary in their analytical and predictive power (Ashby et al., 1989).

Given the rich portfolio of mutagenicity assays that now is available, little payoff is to be expected from further improving mutagenicity tests, unless perhaps human cell systems or human genetic materials can be incorporated into the assays. Attention still needs to be devoted to optimizing the selection of STT batteries for particular purposes, despite their poor performance to date.

STTs need to be developed for nongenetic health effects. These will need to be complemented or even driven by research on biological mechanisms and by other research insights that will make the empirical STTs understandable. In neurotoxicology, for instance, the STTs need to be linked to the understandings of experimental psychology. Many of the tests now widely in use urgently need to be validated as predictors of human-health effects.

As STTs move into much wider routine application, precautions will have to be taken against improper use and interpretation.

Whole-Animal Tests

Animal assays continue to be essential for predictions. They are necessary as guides for regulation, and they generate much useful basic biological information (Tomatis, 1988). Their predictive power continues to need to be evaluated against human health experience.

Like SAR techniques and STTs, whole-animal tests have been better developed for carcinogenicity than for other toxic end points (Huff et al., 1988). They need to be extended to apply to other end points. As this extension is pursued, several workshop efforts should be made to explore how the tests will be interpreted and weighted for decision making. For example, in regulation, it might not be obvious what importance should be accorded reversible alteration of hormone levels, compared with irreversible damage of some other sort. Thinking through these issues might generate implications for the scientific agenda (Kimmel et al., 1989).

Whole-animal tests are burdensome, and consideration should be given to the possibilities for optimizing bioassay test designs (even more than they have been improved over recent years) to obtain reliable information in shorter time and at less expense. For example, in some instances 6-month animal assays might be just as informative as 2-year assays (Kimmel et al., 1989). Therefore, whole-animal tests should be conducted in such a way as to allow systematic observation of multiple health effects within an assay series and perhaps other research areas (such as pharmacokinetics) as well.

The topic of optimal strategic and tactical design of whole-animal tests would be a subject for a future workshop.

Validation

Ultimately, the value of all toxicology testing depends on how well the tests predict human health effects. Therefore, a number of workshop participants indicated that all toxicological methods should be subjected to evaluation with respect to their reproducibility, specificity, sensitivity, and practical applicability as indicators of human health effects. This has been accomplished for some test methods to some extent, and much more is needed.

Despite the value of proxy information, there is no substitute for human data (Erdreich and Burnett, 1985). Therefore, epidemiological research continues to be a central component of health-effects evaluation. Epidemiological research must be supported and nourished as a complement to toxicological research. Coordinated exploration of unusually highly exposed human populations for study to complement particular toxicological research will be essential.

Research Needs and Possibilities

Potency and Severity of Toxic Response

According to several workshop participants, attention needs to be given to the concept and expression of toxic potency and the rate of change of potency relative to dose. The notion of potency has been developed intensively for carcinogenesis assays (DeRosa et al., 1985). Single-value potency estimates are moving toward a more complicated index, as numbers of this kind are being generated by a diversity of assay methods whose biological interpretation must differ. Moreover, such biological variables as sex or metabolism may moderate a molecule's intrinsic potency (such as by attenuating its reactivity with target genetic or other molecules).

Increasingly, toxicological assessments need to recognize that biological effects may be graduated. Moreover, the assessments will have to include evaluations of widely differing severities of outcome (such as behavioral deficits as compared with other neural impairments or nonfatal cancers as compared with fatal cancers).

Exposure Assessment

Risk assessments are compound estimates of adverse health effects given toxicity, toxic potency, and exposure. Effective predictions of human-health risk, therefore, require good estimates of exposure. Improvement of exposure assessments and of the requisite methods should yield major payoffs in risk-assessment quality. Such methods as exposure markers, pollutant modelling, and the like should continue to be refined.

Evaluation of Mixtures

All human exposure to chemicals involves exposure to mixtures. But toxicological research has been pursued largely for single, well-described compounds. As demonstrated by the considerations of the health effects of exposure to gasoline vapors, evaluating mixtures cannot be avoided, and the methods for doing so need to be improved (Vouk et al., 1987).

In general, the evaluation of mixtures (either well-characterized mixtures or such highly variable environmental samples as exhausts, effluents, or soils) tends to be scientifically impure and imprecise. But it can be pragmatically illuminating, helping establish the relative risk of complex chemical hazards in a preliminary way and providing cues for further toxicological investigation.

Research Incentives

There is substantial concern about the problem of making it important and worthwhile for research-based industries to pursue toxicological investigation, exposure-assessment, and related investigations. The topic of incentive systems should be addressed according to several workshop participants in a future workshop, convened by the NRC or another such organization.

Coordination and Planning

From experience in the National Toxicology Program and elsewhere over the past decade, it is clear that it is possible for industrial groups (either commercial groups or such organizations as the Health Effects Institute and the Chemical Industry Institute for Toxicology) to work in coordinated fashion with federal research laboratories. There have been a number of successful consultations and complementary or parallel research programs, although there is some thought that the use of public funds to test proprietary materials already being marketed is inappropriate.

The nomination, selection, and toxicological evaluation of major chemicals for priority testing deserves continual re-examination, as does development of more useful data on industrial production of, use of, release of and potential human exposure to chemicals.

Throughout the workshop, concern was expressed that all approaches be coordinated to gain efficiency. The issues of consistency of approach among research and regulatory agencies also surfaced. A plan to accomplish these objectives did not emerge. According to a number of workshop participants, a formal plan should not be expected; the fields are too dispersed, and considerable disagreement surrounds many of the methods. But leaders and leading institutions should do all they can to coordinate pursuit of these efforts.

REFERENCES

Anderson, E.L., and A.M. Ehrlich. 1985. New risk assessment initiatives in EPA. Toxicol. Ind. Health 1:7-22.

Ashby, J., R.W. Tennant, E. Zeiger, and S. Stasiewicz. 1989. Classification according to chemical structure, mutagenicity to Salmonella and level of carcinogenicity of a further 42 chemicals tested for carcinogenicity by the U.S.National Toxicology Program. Mutat. Res. 223:73-103.

DeRosa, C.T., J.F. Stara, and P.R. Durkin. 1985. Ranking of chemicals based on chronic toxicity data. Toxicol. Ind. Health 1:177-192.

Erdreich, L.S., and C. Burnett. 1985. Improving the use of epidemiologic data in health risk assessment. Toxicol. Ind. Health. 1:65-81.

Hattis, D., and D. Kennedy. 1986. Assessing risks from health hazards: An imperfect science. Technol. Rev. 89:60-71.

Huff, J.E., E.E. McConnell, J.K. Haseman, G.A. Boorman, S.L. Eustis, B.A. Schwetz, G.N. Rao, C.W. Jameson, L.G. Hart, and D.P. Rall. 1988. Carcinogenesis studies: Results of 398 experiments on 104 chemicals from the U.S. National Toxicology Program. Ann. N.Y. Acad. Sci. 534:1-30.

Kimmell, C.A., D.G. Wellington, W. Farland, P. Ross, J.M. Manson, N. Chernoff, J.F. Young, S.G. Selevan, N. Kaplan, C. Chen, L.D. Chitlik, C.L. Siegel-Scott, G. Valaoras, and S. Wells. 1989. Overview of a workshop on quantitative models for developmental toxicity risk assessment. Environ. Health Perspect. 79:209-216.

NRC (National Research Council). 1983. Risk Assessment in the Federal Government: Managing the Process. Washington, D.C.: National Academy Press.

NRC (National Research Council). 1984. Toxicity Testing: Strategies to Determine Needs and Priorities. Washington, D.C.: National Academy Press.

NRC (National Research Council). 1988. Complex Mixtures. Washington, D.C.: National Academy Press.

Tomatis, L. 1988. The contribution of the IARC Monographs Program to the identification of cancer risk factors. Ann. N.Y. Acad. Sci. 534:31-38.

U.S. Department of Health and Human Services. 1986. Determining Risks to Health: Federal Policy and Practice. Dover, Mass.: Auburn House Publishing.

Vouk, V.B., G.R. Butler, A.C. Upton, D.V. Parke, and S.C. Asher, eds. 1987. Methods for Assessing the Effects of Mixtures of Chemicals Scope. New York: John Wiley & Sons.

RELATED PUBLICATIONS

NRC (National Research Council). 1989. Biological Markers in Pulmonary Toxicology. Washington, D.C.: National Academy Press.

NRC (National Research Council). 1988. Complex Mixtures. Washington, D.C.: National Academy Press.

NRC (National Research Council). 1986. Environmental Tobacco Smoke: Measuring Exposures and Assessing Health Effects. Washington, D.C.: National Academy Press.

NRC (National Research Council). 1985. Epidemiology and Air Pollutants. Washington, D.C.: National Academy Press.

NRC (National Research Council). 1984. Toxicity Testing: Strategies to Determine Needs and Priorities. Washington, D.C.: National Academy Press.

NRC (National Research Council). 1983. Risk Assessment in the Federal Government: Managing the Process. Washington, D.C.: National Academy Press.

NRC (National Research Council). 1981. Committee on Indoor Air Pollutants. Washington, D.C.: National Academy Press.

NRC (National Research Council). 1981. Committee on Medical and Biological Effects of Environmental Pollutants. Washington, D.C.: National Academy Press.

Workshop Participants

Research to Improve Predictions of Long-Term Chemical Toxicity
Workshop Participants by Workgroup
Washington, D.C.
December 13-15, 1989

<u>Structure-Activity-Relationship Analyses and Other Correlational Techniques</u>

Dr. Elizabeth Weisburger, Chair, National Cancer Institute (retired), Bethesda, Maryland
Mr. Kurt Enslein, Health Design, Inc., Rochester
Dr. Gilles Klopman, Case Western Reserve University, Cleveland
Dr. Harold Moore, University of California, Irvine
Dr. Ray Tennant, NIEHS, Research Triangle Park
Dr. Yin-tak Woo, EPA, Washington, D.C.

<u>Short-Term In Vitro and In Vivo Tests</u>

Dr. Herbert Rosenkranz, Chair, University of Pittsburgh Graduate School of Public Health, Pittsburgh
Dr. John Frazier, Johns Hopkins University, Baltimore
Dr. Claude Hughes, Duke University Medical Center, Durham
Dr. Marvin Legator, University of Texas Medical Branch, Galveston
Dr. Vernon Ray, Pfizer, Inc., Groton, Connecticut
Dr. Michael Waters, Environmental Protection Agency, Research Triangle Park
Dr. Gary Williams, American Health Foundation, Valhalla, New York
Dr. Errol Zeiger, NIEHS, Research Triangle Park

<u>Whole-Animal Tests</u>

Dr. Michael Gallo, Chair, UMDNJ, R.W. Johnson Medical School, Piscataway
Dr. Deborah Barsotti, Agency for Toxic Substances and Disease Registry, Atlanta
Dr. Thomas Fuhremann, Monsanto Co., St. Louis
Dr. Richard Griesemer, NIEHS, Research Triangle Park
Dr. Emil Pfitzer, Hoffman-LaRoche, Inc., Nutley, New Jersey
Dr. Jerold M. Ward, National Cancer Institute, Frederick, Maryland
Dr. Bernard Weiss, University of Rochester Medical Center, Rochester

Validation

Dr. Roy Albert, Chair, University of Cincinnati Medical Center, Cincinnati
Dr. David Hoel, NIEHS, Research Triangle Park
Dr. Kim Hooper, State of California Department of Health, Berkeley
Dr. Renate Kimbrough, Environmental Protection Agency, Washington, D.C.
Dr. Peter Magee, Temple University School of Medicine, Philadelphia
Dr. Barry Margolin, University of North Carolina, Chapel Hill
Dr. John Mulvihill, University of Pittsburgh Graduate School of Public Health, Pittsburgh
Dr. David Peakall, Canadian Wildlife Service, Ottawa

Strategy, Policy, and Resource Considerations

Dr. William Lowrance, Chair, Rockefeller University, New York
Dr. John Andrews, ATSDR, Atlanta
Dr. Irv Baumel, Health Effects Research Division, U.S. Army, Biomedical R&D Lab, Frederick, Maryland
Dr. Irina Cech, U.S. House of Representatives, Energy and Commerce Committee, Washington, D.C.
Dr. William Farland, Environmental Protection Agency, Washington, D.C.
Dr. James Huff, NIEHS, Research Triangle Park
Dr. Jack Moore, Institute for Evaluating Health Risks, Irvine, California
Dr. Andrew Sivak, Health Effects Institute, Cambridge
Dr. James Wilson, American Industrial Health Council, Washington, D.C.
Dr. John Young, Hampshire Research Associates, Inc., Alexandria, Virginia

Observers

Dr. Devra Lee Davis, Scholar in Residence, National Research Council, Washington, D.C.
Mr. Carl Gerber, Environmental Protection Agency, Washington, D.C.
Dr. Sid Siegel, National Library of Medicine, Bethesda

Appendix D

Research Needs in Anticipation Of Future Environmental Problems

A Workshop Report

Summary

Society frequently reacts to environmental problems only after they become public crises, then wishes that timely research had helped either to anticipate the crises or to provide means to deal with them. The absence or inadequacy of relevant scientific knowledge and understanding frequently makes it difficult to generate rational environmental policy to deal with problems as they arise. The best that can usually be done is to base responses on imperfect information and what seem to be reasonable hypotheses. It is, of course, impossible to anticipate all problems. The key question is which areas of research should be pursued now so that after 5, 10, 20 or more years, society will be in a better position to identify and respond to serious environmental problems. This is what the committee calls "anticipatory research."

This report summarizes a workshop on anticipatory research held in Woods Hole, MA, June 14-15, 1989, to conceive and define research programs not already under way that would help the federal government, states, industry, universities, and other research organizations in anticipating environmental problems and providing a base of scientific knowledge in time to shape a response. Workshop participants included environmental scientists and engineers who were asked to focus on responses to the following questions:

1. What should be the shape of a long-term research program to cope with these problems?
2. How can arising problems be identified in an ongoing way?
3. What potential problems can be identified now?

Discussion of the responses to these questions at the workshop led participants to propose the following responses:

- *Management:* The overall goal of a more fully developed capability to anticipate and respond effectively to future environmental problems is to improve our ability to manage the environment. The overall objective of improving the capability to manage the environment will be well served through improving the heterogeneity of ideas, approaches, and activities in environmental research; using and balancing holistic and mechanistic studies; and using and balancing system-level and process-level investigations.
- *Institutional arrangements:* Several alternative institutional arrangements should be investigated, including working within existing institutions as well as establishing new ones. Whatever alternative is selected, anticipating future environmental problems, identifying anticipatory and exploratory research needs, conducting the requisite research and acting on the results have to be given a clear, continuing, and unambiguously high priority. Environmental organizations should consider establishing new units charged with an anticipatory, exploratory responsibility. Past attempts such as EPA's Washington Environmental Research Center should be examined as possible models, as should programs in other environmental organizations, in the United States and in other countries. Whatever the organizational structures selected, these units should be inter- and multidisciplinary.
- *Environmental personnel:* Government and industry have an obligation to see that the environmental sciences have sufficient funding and visibility to ensure adequate recruitment of bright, capable individuals. Ways to effectively use and

retain successful, experienced scientists should also be implemented. Funding organizations should recognize that researchers need to have flexibility and that some portion of each grant or contract should be devoted to development of new ideas.

• *Understanding past environmental problems:* A preliminary step in designing an exploratory research program to anticipate environmental problems should be studies of the past to determine the reasons that environmental problems often go unnoticed and how early identification and resolution can be achieved. Understanding past environmental problems can be improved through: an analysis of case histories and synthesis of past efforts to manage environmental problems; study of the relation between the contribution of science and other factors to successful environmental management; scrutiny of research programs at existing academic, government, and independent research institutes to identify those factors that have contributed to highly successful research programs; evaluation of the effectiveness, advantages, and disadvantages of programs carried out by individual investigators, research teams, and centers to identify the relative advantages and disadvantages of alternative approaches; establishment of an unambiguous high priority and long-term commitment to environmental studies; and international cooperation to exchange ideas, expertise, and experience.

• *Monitoring programs:* Any sound environmental research program aimed at achieving a better understanding of our environment in the future should have two basic objectives: to know what the situation is now (i.e., to establish baselines), and to recognize and follow changes, to provide warning and to determine the seriousness of emerging threats. A critical question is what to monitor and where. Several key elements of a monitoring program are: a multidisciplinary advisory panel; inventories of resources, emissions/discharges, chemical use, and other vital statistics of societal and industrial activity; social, economic, and technological trend data; and active research, modeling, and diagnostic programs to assess environmental conditions now and in the future.

A program to manage physical phenomena should establish baseline conditions in the biotic and abiotic components of the environment; quantify trends in pollutant concentrations in the environment and in the condition of ecological resources; define the magnitude, rate, extent, and location of changes; and provide data to allow relationships between anthropogenically induced changes and alterations in the quality and quantity of biotic systems to be determined.

Three key social and economic areas that need to be monitored are environmental consequences of human behavior, impact of management and regulatory initiatives, and the relations between societal values and environmental expectations. The ways public values and behavior interact with environmental problems are poorly understood. A series of studies exploring how public values and behavior might be expected to change over the next several decades and the implications of such changes on the perception of environmental problems and the constraints on the potential effectiveness of alternative management strategies should be undertaken.

• *New technology studies:* New technologies may involve either new materials or new processes whose environmental impact is unknown. In such cases, the introduction of the major new technology should be preceded by environmental and societal impact analysis of the process, the manufacturing materials and the products and byproducts. However, the hazard of a new technology should be compared not with an arbitrary standard, but to the technology currently in place. In other words, the comparison made should be relative rather than absolute. Many industries require environmental monitoring, yet the financial resources to perform adequate monitoring may be lacking. It is important to expand the monitoring to the entire life cycle of each product, in terms of cradle-to-grave mass balances of products and byproducts. A research program on new technology should ask what knowledge is needed to reduce the risks of introducing new technologies to an acceptable level, what critical knowledge regarding environmental impact the history of the introduction of technologies shows to be necessary, and whether there are general lessons or whether each case must be analyzed sui generis. Two emerging areas where consideration of environmental consequences before implementation might alleviate problems in the future are renewable energy and advanced materials.

Workshop participants came to the broad conclusion that, if future environmental problems are to be met, a broad based, long-term research program is needed in addition to the short-term research and testing programs that provide the scientific foundations for current regulatory problems. Such a program will require well-designed environmental monitoring to establish baselines from which environmental changes can be measured and then followed. It will need ongoing studies of the changing social situation, the public's expectations for environmental management, and the outcomes of environmental interventions. New technologies need to be evaluated from two points of view: what new environmental problems they might generate and how they might contribute to the alleviation of environmental difficulties.

Research Needs in Anticipation of Future Environmental Problems

INTRODUCTION

This report summarizes a workshop on anticipatory research held in Woods Hole, MA, June 14-15, 1989, to conceive and define research programs not already under way that would help the federal government, states, industry, universities, and other research organizations in anticipating environmental problems and providing a base of scientific knowledge in time to shape a response.

Society frequently finds itself reacting to environmental problems when they become public crises and wishing that timely research had helped either to anticipate the crises or to provide means to deal with them. Acidic deposition, biomagnification of DDT, asbestos fibers, and clean-up standards for ground water and soil, all are examples of problems for which we might have been better prepared. The absence or inadequacy of relevant scientific knowledge and understanding frequently makes it difficult to generate rational environmental policy to deal with problems as they arise.

In some cases, testing programs or other short-term investigations can provide the necessary information. However, many problems, such as devising generic rules for identifying substances involved in the carcinogenic process or estimating the ecosystem impacts of alternative energy scenarios, are not amenable to direct, short-term attack. Several workshop participants believe that what is needed in many cases is a quantitative understanding of the basic processes involved. However, if basic research programs are required, one needs to decide what areas should be studied and what environmental questions are likely to arise. Otherwise there would be no limit to the exploratory programs that might be undertaken.

Because current problems require immediate action, the best that can usually be done is to base responses on imperfect information and what seem to be reasonable hypotheses. Yet there are reasons to believe that many of the current estimates of risk, and hence of the urgency of the requirement for particular actions, may be in error by factors as great as a million and, therefore, that the funds spent for alleviating environmental risk may be grossly misallocated. There is nothing that can be done about that now, but if appropriate research had been carried out for the past 20 years or more, we would be in a better position today to deal with many of these problems.

It is, of course, impossible to anticipate all problems. However, future problems can sometimes be discerned through informed speculation based on current trends (e.g., health statistics, population structure, building stock and infrastructure, and new industrial technologies). The question is what areas of research to pursue now so that after 5, 10, 20, or more years, society will be in a better position to identify and respond to serious environmental problems.

Past Anticipatory Research Efforts

Past attempts to improve our ability to anticipate and respond to environmental problems, beginning with the 1965 report, *Restoring the Quality of the Environment*, by the President's Science Advisory Committee, have had limited success. For example, in 1971, the U.S. Environmental Protection Agency (EPA) established the Washington Environmental Research Center to develop a strategic environmental modeling and socioeconomic studies capability. For a variety of reasons, the center was abolished in the mid-1970s.

The next attempt by EPA began in 1978, when, with congressional support, a strategic anticipatory analysis and exploratory research program was developed in the Office of Research and Development. The Office of Exploratory Research (OER) was created, comprising the Office of Strategic Assessment and Special Studies (OSASS) and the Office of Research Grants and Contracts (ORGC). Both offices participated in the EPA's research planning and budget-making processes. OER still exists, but only the grants and centers programs remain; OSASS was abolished in the mid-1980s.

The National Institute of Environmental Health Sciences (NIEHS) is the only federal institution directed to focus on the underlying science relating to human health and environmental interactions. In addition to conducting basic and applied research studies in its intramural programs, it also has provided continuing support to university-based scientists in its extramural program. Of particular note here are the multidisciplinary "center grants" supporting broad-based environmental health programs with a variety of emphases. These centers are one of the leading sources for well-trained scientists (at pre- and postdoctoral levels) in the environmental health sciences.

Workshop Focus

Workshop participants were asked to focus on responses to the following questions:

1. What should be the shape of a long-term research program to cope with these problems?
2. How can arising problems be identified in an ongoing way?
3. What potential problems can be identified now?

The concept of exploratory research is large and heterogeneous, and an exhaustive list and analysis of the possible structure and research needs is impossible. Therefore, workshop participants focused on a few areas that appeared to offer potential for exploratory research: monitoring, impact of future changes in energy and technology, impact of societal and economic trends, and the process of designing exploratory research programs.

IDENTIFICATION OF ARISING PROBLEMS

Any sound environmental program for the future needs to establish baselines for the current situations, recognize and follow changes, provide warning, and determine the seriousness of emerging threats. This kind of program requires the monitoring of physical and social phenomena.

A critical question is what to monitor and where. To help ensure that adequate attention and oversight is given to these important questions, workshop participants included the following requirements for the monitoring program:

- Multidisciplinary advisory panel including a diversity of natural, social, and health scientists.
- Inventories of resources, emissions and discharges, chemical use, and other vital statistics of societal and industrial activity.
- Close contact with a social, economic, technological trends program aimed at anticipating future issues that should be incorporated into the program as well as provide feedback on current programs.
- Active research, modeling, and diagnostic programs to assist in the selection of the best indicators of environmental conditions now and in the future and to provide the necessary interpretation of relevant trends, especially with regard to ecosystems.
- Biological and chemical methods and instrumentation development program to advance our ability to make needed measurements efficiently and accurately.
- Extensive quality assurance programs for data collection and management to ensure comparability, accuracy, precision, and, ultimately, usefulness over the long term.
- Reporting element that regularly conveys the findings for both the scientific community and the decision makers.
- Accessible database for all who wish to assess and explore the information, refine the interpretations, or gather insights for managing the environment.
- Close coordination among the programs of the various agencies that may be involved.

Monitoring Physical Phenomena

Few programs are adequately designed to examine the status and trends in the physical environment other than the global-scale programs of the National Aeronautics and Space Administration, National Oceanic and Atmospheric Administration, National Science Foundation, and U.S. Geological Survey. These agencies monitor the concentrations of the "greenhouse gases," such as CO_2, N_2O, CH_4, and the chlorofluorocarbons that are responsible for the drop in stratospheric ozone concentrations. There are also programs to monitor global temperatures, sea levels, and changing ocean currents. All of these monitoring programs are part of large-scale, ongoing research programs aimed at understanding the evolving status of the earth.

However, regional and local environmental problems raise difficult questions regarding what should be monitored and in what detail. Such questions as how point measurements of air or water quality should be combined to draw more general conclusions are difficult theoretical questions. As a consequence, we are rarely able to address emerging environmental problems with the certainty required for decision makers to take action. The situation arises because our monitoring efforts are primarily compliance oriented and hence designed to determine whether criteria, standards, or permissible levels are being exceeded.

Therefore, workshop participants suggested that an environmental monitoring program should be designed and implemented in areas other than those included in current programs of such organizations as NASA, NOAA, NSF, and the USGS. The program should have an international focus and be designed to accomplish the following:

- Establish baseline conditions in the biotic and abiotic components of the environment.
- Quantify trends in pollutant concentrations in the environment and in the condition of ecological resources.
- Define the magnitude, rate, extent, and location of changes.
- Provide data to determine the relationships between anthropogenically induced changes and alterations in the quality and quantity of biotic systems.

Because a primary goal of a monitoring program is to establish trends, long-term funding is important if that program is to be successful. An example of a successful long-term study is the "Six Cities" study, in which air quality and the health status of a large cohort of children and adults were monitored for more than a decade in six cities with varying pollution levels. Only now are important conclusions emerging.

Detecting long-term environmental trends or identifying damage requires that statistical designs receive particular emphasis in the new monitoring programs. Issues of scale, measurement frequency, variability within and among systems, and desired precision, to mention a few considerations, all need to be evaluated with the aim of improving efficiency and consequently reducing costs of monitoring. In addition, as global, national, regional, subregional, and local environmental problems need to be considered, coordinated designs for monitoring programs are warranted, because they provide flexibility in making appropriate measurements at correspondingly appropriate scales while achieving maximum aggregation and disaggregation options for exploratory analyses of the data collected.

Societal and Economic Monitoring

The design of long-term exploratory research programs must take into account the societal and economic trends that will generate new environmental problems or change the character of problems that are already apparent. Workshop participants identified three social and economic areas that need to be monitored:

- Environmental consequences of human behavior.
- Management and regulatory initiatives.
- Relationship between societal values and environmental expectations.

Currently, the process of societal values and behavior interacting with the environment to generate health and ecological impacts is poorly understood. Some businesses resist changing habitual production patterns even when shown that different options would be both economically and environmentally beneficial. Homeowners may actively oppose the construction of waste-treatment facilities near their homes, yet show no interest in inexpensively testing their own homes for radon, which potentially poses a greater hazard. Periodic scares—medical wastes on beaches, tainted Chilean grapes, or EDB in foodstuffs—generate public calls for action, while other more serious risks produce only limited responses. Many public policy interventions intended to achieve environmental protection miscalculate human behavior and its environmental consequences.

Environmental decisions are typically based on a narrow set of considerations that are important to society. Thus, the value of protective actions is customarily measured by reference to health and injury effects, direct economic impacts, and effects on a small number of specific, measurable environmental indicators. This range of effects omits many issues relevant to public values and concerns and thus contributes to the seemingly perplexing divergence between the evaluation of the importance of problems by public agencies and the public's responses. The difficulty is that even if risks to health and ecology were reduced at a feasible cost, people still would not be well served if it were done at the expense of the happiness derived from esthetics, meaningful work, or harmony with nature. Research is needed to develop more comprehensive methodologies for assessing the human values resulting from environmentally protective actions.

The monitoring of each of these areas—environmental consequences of human behavior, management and regulatory initiatives, and societal values—can be done through forecasting by extrapolating from past and current trends and through constructing goal and surprise-oriented scenarios.

Forecasting Long-Term Socioeconomic Trends

According to workshop participants, forecasting and analysis of basic long-term social and economic trends can provide important clues and insights into potential future environmental problems. Some trends meriting analysis are demographic changes, changes in gender ratios in the work place, changing consumption patterns and uses of new products, shifts in industrial patterns and technologies, and changes in urbanization patterns.

Careful analyses of such trends can improve the capability to anticipate changing exposure patterns, changing vulnerability to various natural and technological hazards, new environmental stresses (e.g., changing land use and emergence of megacities), and new or underestimated hazards. Research should be done to forecast key social and economic trends over the near and the long term. The emphasis should be on assembling data and projections now scattered through a variety of institutions, exploiting them to forecast key social and economic trends pertinent to environmental impacts, and identifying major environmental problems that may occur. Consideration should be given to how potential problems can best be characterized.

Goal- and Surprise-Oriented Scenario Construction

While extrapolation of past and current trends should improve our analytical ability to anticipate future environmental problems, workshop participants believe that this approach should be complemented by a very different type of analytical thinking, namely scenario construction. Scenario construction can make several important contributions to society's anticipatory capabilities. It directly combats the mind sets referred to above and challenges linear types of thinking. By encouraging anticipatory thinking, scenario construction contributes to an improved early-warning capability. It generally enhances rapid response and managerial adaptiveness in the face of surprise.

Workshop participants recommended that research should be undertaken and maintained to support the following two types of scenario construction:

- *Type 1*: Goal-oriented. Here the scenario begins with a postulated goal—for example, a low-energy society in 2020, a fossil-fuel phase-down, and a global strategy of sustainable development. The scenario construction would then address technologies and social structures that would have to be in place for the goal to be realized;

alternative pathways for getting there; and environmental, resource, and social implications for each pathway.

- *Type 2*: Surprise-oriented. This analysis begins by projecting a future state of society that may reasonably be expected to exist at a specified time. The analysis then treats the major attributes of that societal state, pathways for getting there, possible surprises or disjunctures that may occur along each pathway, the environmental consequences associated with the pathway and surprises, and the likely adequacy of societal warning systems and coping abilities.

Comparative Experience

Better use should be made of comparative experience. European nations, for instance, have adopted some approaches different from those used in the United States. The nations of Europe and elsewhere have also developed experience in dealing with the environmental problems associated with high population densities and urban and industrial activities, and in some cases have developed innovative and different approaches. Research should be designed to identify which approaches are most effective and how the transferability of success elsewhere can be assessed and facilitated.

Further, workshop participants suggested that with increasing recognition of the impact of transnational issues, the scale of multinational corporative actions, and the diversity of cultural settings and values, research is needed to clarify the management issues posed by an increasingly interdependent world, and the institutional and regulatory developments needed to fill current voids.

Specific Areas That Require Monitoring

Workshop participants identified specific areas in the physical and societal environment that require monitoring.

Biotic Monitoring

- Integration and critical analysis. Monitoring data on the structure and function of terrestrial and aquatic ecosystems have the potential to identify emerging environmental issues and problems. Organisms in the environment can serve as integrators of environmental stress. Changes in their physiology, biochemistry, species richness, diversity, etc., may be early warnings of emerging environmental problems. Biological monitoring of the environment can be conducted at various levels of ecological organization. Frequently monitored levels include organism, population, community, and ecosystem. A current critical research need is to develop clear ecological assessment end points at each of these levels of organization to be used in biomonitoring of the environment. Assessment end points are formal expressions of the actual environmental values that are to be protected. Once these assessment end points are established, measurement end points that can be implemented efficiently can become part of a biomonitoring program.

- Biological markers. Biochemical, physiological, and morphological manifestation of anthropogenic stress—"biological markers"—have the potential to provide early indications of adverse effects to organisms. Research to relate these biological marker changes to effects on survival, growth, and reproduction is currently being conducted in a number of laboratories. It is recommended that biological markers be employed in the monitoring program where appropriate and that developments in this rapidly developing field be closely followed.

- Regional, national, global levels. Most biomonitoring programs have been designed to assess local environmental concerns, but regional, national, and global scales of concern must also be addressed. New assessment and measurement approaches designed to address environmental issues at these levels may need to be developed. Establishment of a biomonitoring program to identify emerging environmental problems requires more than just collecting data and plotting trends. It requires the integration of physical, chemical, biological, and demographic data. The use of data management systems such as Geographic Information Systems (GIS) has the potential to assist the environmental analyst in this complex job. Thus, it is recommended that high priority be given to research on the application of GIS technology to monitoring data. GIS has particular value at the regional, national, and international scales of environmental concern.

- Biotechnology. The increased interest in biotechnology and the resulting production and use of genetically altered organisms have led to questions about their possible harm to the environment. Although such organisms are unlikely to be ecologically successful outside the habitat in which they are introduced, we must nonetheless develop a fundamental understanding of their survival and dispersion in various ecosystems and devise techniques to monitor their fate in the environment.

To achieve these goals, appropriate research must be undertaken.

Abiotic Monitoring

- Natural waters. A considerable increase in the scientific data base on the organic and inorganic composition of natural water systems could result from the deliberate substitution of the question "What compounds are present?" for the question employed in priority pollutant analysis, "Is this compound present?" In almost all chemical monitoring programs conducted today, chemicals that are thought to be important are quantified. Whether a chemical is investigated often is related to whether it has caused a problem previously. The number of compounds tested for has increased from approximately 10 halogenated organic pesticides in the 1960s to more than 100 substances on the priority pollutant list used today. In these programs, only compounds on the list generate a signal, which is fed back for some informational or regulatory purpose. Compliance monitoring or NPDES permit monitoring are examples of this approach.

The analytical methodologies developed for these monitoring programs often intentionally eliminate or ignore compounds not on selected lists. Most of the information generated by the analytical instruments involved, such as gas chromatographs and mass spectrometers, is discarded. With the development of computerized analytical data systems however, all analytical signals can be collected through A to D converters and stored on disk or tape.

Standardizing methodologies allows comparison of data generated over time at different laboratories by different individuals. Software can be generated to investigate total data sets to determine if new compounds or new classes of compounds are entering the system. Temporal or aerial concentration trends can be evaluated. Data systems from various laboratories can be directly linked together. The chemical monitoring data base could increase exponentially. Because the extent of such a system would be limited by cost, the sampling scheme needs careful design.

If the intent of the program is to test hypotheses or to follow trends, then the analytical methodologies should be such that the accuracy of the data will be maximized. However, to maximize the probability of detecting or observing a new compound or class of chemicals, a broad qualitative scheme at the sacrifice of some accuracy may be in order.

Societal Monitoring

- Characterization of population. The potential character of an aging global population over the next 50 years and the magnitude of the effects on health and environment should be analyzed.
- Analysis of changing values and behavior. The possibility of values and behavior changing over the next several decades should be explored, and the implications of such changes on the perception of environmental problems and the effectiveness of alternative management strategies should be studied.
- Assessment of the nation's capability to deal with major environmental problems in terms of behavioral and value constraints. This assessment should specifically identify human values or behaviors that are most amenable to changes designed to increase human or environmental protection and those that are most resistant to change. It should ask whether it is reasonable to believe that such changes can occur and how such changes may best be pursued.
- Comprehensive analysis of management intervention outcomes. The analyses performed to develop management strategies and subsequently to evaluate the strategies often lack the scope necessary to foresee all or even the most relevant effects of environmental management on society, including outcomes such as substitution of one hazard for another;

secondary and tertiary effects of decisions; new hazard creation; or social, technological, or economic changes. Improved methodologies should be developed to predict outcomes of intervention better.

- Understanding of management thinking. All managers and institutions are prisoners, to some extent, of ingrained ways of thinking about and responding to problems. A current example is the tendency in EPA to assume that point-source pollution should be dealt with by regulation and nonpoint source pollution by incentives or voluntary compliance. Meanwhile, the key environmental problem has shifted from regulation to incentives. Research is needed to assess the frameworks that now exists among corporate and governmental environmental managers, how these frameworks affects the abilities of our institutions to anticipate or respond in timely fashion to new environmental problems, which frameworks have an adverse impact on the environment, and how these frameworks might be changed.
- Extending valuation analysis. New valuation estimation methods should be developed that treat qualitative aspects of environmental hazards, secondary and tertiary impacts of regulatory action, cumulative and long-term effects of regulatory action, and the broader domain of moral discourse concerning environmental values.
- Ongoing and systematic appraisal of changing values and concerns. A rigorous program of monitoring and appraisal would require the best of social science expertise. Specifically, it should extend well beyond polling and survey efforts to employ a diversity of approaches and theories, as well as sustained secondary analyses of data gathered. The design of such a program should be participatory in nature and broadly peer reviewed.

Developing an ability to discriminate between types of public response, if properly done, should provide an enhanced ability to discriminate between two different kinds of public reactions: Those arising from misinformation, issue dramatization, and extensive media coverage, and those rooted in basic and enduring public values, societal or group goals, or specific public knowledge.

Information Dissemination

It is important to identify how best to inform various national and international institutions of the results of the overall forecasting and assessment efforts referred to above. Of paramount importance to a monitoring program designed to anticipate future environmental problems is the integration and evaluation of the data generated. At this time, no framework or mechanism is in place to allow such considerations. Therefore, workshop participants recommended that a data management system be developed with the responsibility to integrate various data sets.

TWO POTENTIAL PROBLEMS: ADVANCED MATERIALS AND RENEWABLE ENERGY

The introduction of any major new technology should be preceded by environmental and societal impact analysis of the process, the manufacturing materials, and the products and byproducts, according to workshop participants. Microelectronics industries, for instance, use many hundreds of materials that have largely unknown consequences on human health and environment. Organometallic compounds such as trimethyl arsenic; semiconductor substrates, such as gallium arsenide; and metal alloys used in superconductors are several examples of materials used in new products.

Many industries require environmental monitoring, yet the financial resources to perform adequate monitoring of all industries may be lacking. It is important to expand the monitoring to the entire life cycle of each product, in terms of cradle-to-grave mass balances of products and byproducts. How can the innovator of a new technology participate optimally in environmental protection? To what extent should public support be anticipated by industry? What mix of public and private support and participation in environmental protection would be optimal, given both environmental protection needs and the importance of incentives for technological research and development? What incentives will foster research and development of technologies with acceptable environmental impact?

All of these questions could be answered better after suitable research. A research program should ask what knowledge is needed to reduce the risks of introducing new technologies to an acceptable level, what critical knowledge the history of introducing technologies shows to be necessary regarding environmental impact, and whether there are general lessons or whether each case must be analyzed sui generis.

Workshop participants identified two emerging areas—renewable energy and advanced materials—where consideration of environmental consequences before implementation might alleviate problems in the future.

Renewable Energy

The possibilities of significant increases in the use of renewable energy sources have been expanded through technology, for example, more efficient transmission of hydroelectric energy if superconductivity became commercially feasible. However, too little attention has been given to the potential negative consequences of significant increases in the use of each of the potential energy sources. Therefore, the examination of an optimistic but credible scenario of a significantly higher reliance on renewable energy resources for, say, the year 2020, should be undertaken as a research program to focus on the consequences of expanding such alternatives as hydroelectric power (e.g., silting, salinity, and other problems associated with dams), agro-energy production (e.g., land scarcity and pesticide pollution potentially arising from sugar-cane-based ethanol production), and a methane/methanol-based system. More generally, plausible future scenarios of different kinds of energy generation and delivery systems should be assessed to determine the relevant byproducts and to evaluate the consequences of these byproducts.

Part of the associated research effort could be to study the byproducts of current applications of these systems (for example, the level of leakages in methane and methanol systems) and to undertake applied research to find ways to reduce the total inputs necessary for these various energy systems, both to economize and to reduce harmful byproducts.

Even though most engineering studies of systems, whether or not they involve new technologies, presume that safeguard systems will operate as planned and that control procedures will be followed faithfully, experiences such as Three Mile Island, Chernobyl, and Bhopal indicate that accurate forecasts of environmental hazards ought to incorporate some scenarios of suboptimal performance, such as accidents.

For existing technologies, such as nuclear power plants, a record on the frequency of errors can provide a basis for forecasting, but for new technologies, the likelihood of environmental risks must be based on analysis of the systems and modeling of the possible consequences such as human error, equipment failure, and sabotage. Many issues are involved in anticipatory research, as opposed to a single risk assessment, because the problems change easily over time due to new technologies and the effects of increased use.

Advanced Materials

New materials should be designed in such a way that desirable commercial and economic properties are combined with attractive environmental properties such as recyclability, reusability, low energy use in production and recycling, inertness, and low toxicity. However, some new products and materials introduced into the marketplace, such as composite materials, zinc-coated steel, microchips, disposable diapers, and multi-layered plastic packaging materials, show that these objectives have not been achieved.

Biotechnology offers, in principle, the possibility of manufacturing products that come closest to fulfilling these objectives. It will have a significant impact on agriculture, human health, and chemical manufacturing as a result of our new-found ability to combine genetic material from organisms in widely divergent taxonomic groups in precise, rapid ways.

Biotechnology, like any other technology, also has potential negative aspects that must be taken into consideration. Environmental application of genetically modified organisms is by no means a new human endeavor. We already have a large body of data on environmental releases of modified and unmodified organisms produced by selective breeding; for the most part, these releases have posed few environmental problems. In the case of genetically engineered organisms, the risks can be minimized by careful, responsible design and use of organisms after appropriate research.

Some specific deleterious consequences that may result from releasing transgenic organisms include the following:

- Transfer of new genetic material from the engineered organism to a nontarget species. For example, engineered traits may move from engineered crop plants to their wild, weedy relatives. Depending on the trait, increased competitiveness, subsequent range expansions, and future weed problems may result.
- "Cascade" effects resulting in the loss of beneficial species that are ecologically distant from the transgenic organism.
- Disruption of community structures and dynamics as populations of competitors, predators, parasites, and hosts shift.
- Unintentional broadening of host ranges of pathogens to include beneficial species.
- Possible narrowing of the genetic base of crop plants, thus creating unforeseen pest problems.

Workshop participants noted that renewable energy and advanced materials are examples of areas in which ongoing research is needed so that potentially negative consequences of large scale use of biotechnology implementation can be determined.

DESIGN FOR AN EXPLORATORY RESEARCH PROGRAM

To some extent, the reasons for the poor performance of the scientific community in anticipating environmental problems can be identified, according to workshop participants. Some of these reasons are discussed below.

- An adequately integrated conceptual theory or framework for how environmental systems function is lacking. In the absence of such a framework to guide research and set priorities, much of what has been learned about natural systems has come about incidentally from studying specific problems as they arise. For example, the fundamental concepts of nutrient cycling and energy flow in ecosystems were considerably advanced by studies in the 1960s and 1970s on the movement of radioactive waste products through various geological configurations.
- Closely associated with the absence of a framework for environmental sciences has been the lack of a singular scientific discipline to embrace and integrate the knowledge needed to solve environmental problems. Environmental sciences and problem solving are typically taught and practiced as subspecialties of individual disciplines such as chemistry, biology, engineering, earth sciences, economics, and policy analysis. Thus, individuals expert in one discipline but unschooled in the basics of another area are often ill prepared to anticipate environmental problems that call for multidisciplinary skills.
- Within the organizations with environmental responsibilities (with the exception of the NIEHS and those dealing with global phenomena) the emphasis is primarily, if not exclusively, on the near term. Because managers' performances are evaluated on the basis of how well they deal with existing environmental problems, a disproportionately large amount of resources is invested in short-term problem solving. This emphasis contributes to the inability to anticipate, identify, and act decisively in a scientifically informed manner on environmental problems. No systematic program to assemble ideas and evaluate new, emerging, and escalating environmental problems has been sustained. Thus, an exploratory research program should have an integrative framework, and be interdisciplinary, coordinated, and long term.

An example of the difficulties described above is ecotoxicology, the study of the fate of chemicals released to the environment and their effects on individuals, species, and ecosystems. Ecotoxicology is a young science that has not been recognized by funding agencies; thus, research priorities are still identified within traditional disciplines. A fundamental understanding of environmental sciences must be fostered and activities that improve the use of fundamental mathematical, physical, chemical, and biological processes should be emphasized in an ecotoxicological context.

In the United States, as well as in other countries, responsibility for ecotoxicological issues is split and fragmented to such an extent that communication between closely related areas of responsibility is poor or completely absent. This situation results in haphazard approaches to ecotoxicological problems that are slow and unpredictable and lacking in clear and global objectives. One agency may support short-term ecotoxicological research to solve a specific environmental problem while another federal agency may support basic research in core disciplines with the expectation that this knowledge will eventually be applied to environmental problems. No federal agency has historically supported basic research in the environmental sciences to anticipate ecotoxicological problems.

The problems resulting from this lack of theoretical and disciplinary guidance are compounded by the fragmentation of environmental science and engineering responsibilities among federal agencies, institutes, and national laboratories. Although mechanisms exist that are intended to ensure coordination and cooperation, neither an overall federal environmental research and development plan nor a clearly articulated division of responsibilities exist.

Objectives of an Anticipatory Research Program

The overall goal is to improve our ability to manage the environment by developing capability to anticipate and respond effectively to environmental problems. That is, the goal is to develop the fundamental understanding and the analytical methods and tools needed to achieve designated environmental objectives. The following goals were developed by workshop participants to achieve the overall objective of improving our capability to manage the environment:

- Improve the heterogeneity of ideas, approaches, and activities in environmental research. Enhanced communications and collaboration among environmental scientists of different disciplines will promote innovative thinking and development of the truly multidisciplinary approaches and understanding needed to solve local as well as global environmental problems. This heterogeneity not only has the advantages of hybrid vigor and the enhanced creativity fostered by nontraditional thinking but also encourages use of the collective expertise of several disciplines to yield the best solutions to problems.
- Incorporate holistic and mechanistic studies to contribute information for resolving environmental problems. Mechanistic studies provide precise answers to narrowly defined studies through, for example, controlled experimentation. Holistic studies, by studying the actual conditions in the real world, for example, incorporate a broad and realistic range of factors but at a sacrifice in precision. The ability to generalize from a particular instance or to predict on the basis of partial information is directly dependent on our fundamental understanding of processes. This understanding may be molecular (e.g., the effect of an organic chemical on DNA) or global (e.g., the processes that control the carbon cycle). A fundamental understanding of most environmental processes is developing slowly, chiefly as a byproduct of other research rather than as a defined goal in itself.
- Use system-level and process-level investigations. This use might be established by encouraging collaboration, for example, between people modeling environmental systems (e.g., global atmospheric conditions) and others interested in measuring chemicals in air as indicators of particular reaction processes. Thus, resources should be invested in research involving both mechanistic understanding and field measurements to test modeling results.

Implementation

Four key ingredients were identified to implement an improved long-term environmental R&D program: an effective institutional arrangement; highly qualified environmental scientists and engineers; adequate, dedicated, and continued funding; and the means to assemble promising ideas on a continuing basis.

Institutional Arrangements

Alternative institutional arrangements should be investigated, including working within existing institutions as well as establishing new ones, according to workshop participants. Whatever alternative is selected, anticipating future environmental problems, identifying anticipatory and exploratory research needs, conducting the requisite research, and acting on the results have to be given a continuing and unambiguous high priority. Personnel and financial resources need to be committed, and the performance of environmental mangers should be evaluated on the basis of long-term implications and consequences as well as on the basis of near-term results.

Environmental organizations should consider establishing new units charged with an anticipatory, exploratory responsibility. Past attempts, such as EPA's Washington Environmental Research Center, OER, OSASS, and ORGC, should be examined as possible models, as should programs in other environmental organizations, in the United States and in other countries. Whatever the organizational structures selected, these units should be inter- and multidisciplinary to improve interaction and exchange of ideas among critical disciplines. A more fundamental break with the past should also be considered. For example, a national institute patterned after the National Institutes of Health (NIH) should be examined. In contrast to EPA's programs, which support short-term regulatory objectives, NIH has a research mission dedicated to high-quality, long-term, fundamental research in its own laboratories. Furthermore, NIH provides stable research funding to public and private institutions and provides visible and attractive careers to talented young people (beginning with an excellent pre- and postdoctoral fellowship program).

Environmental Scientists and Engineers

The exploratory research program in EPA should be designed and implemented to attract young, bright professionals as well as accomplished, experienced researchers. A shortage of first-rate talent in the environmental field apparently exists. This shortage is predictable in that environmental sciences may not be perceived as an attractive career by the best and brightest. Many talented undergraduate chemists, biologists, and engineers would be interested in environmental careers if they perceived (1) the intellectual and scientific challenges in the environmental sciences, (2) the opportunity to have a significant effect on protecting the environment and ameliorating environmental contamination, and (3) the opportunities for career development.

Existing environmental programs have an unfulfilled responsibility to give visibility to environmental careers and to avail the profession of a larger diversified applicant pool. A fellowship program to support graduate students and postdoctoral fellows would make an important contribution to this goal.

Overall, workshop participants concluded that government and industry have an obligation to see that the environmental sciences have sufficient funding and visibility to ensure adequate recruitment of bright, capable individuals. Ways to use and retain successful, experienced scientists should also be implemented. Funding organizations should recognize that researchers need to have flexibility and that some portion of each grant or contract should be devoted to developing of new ideas.

Environmental Research and Development Funding

In the discussion on funding, participants agreed that care is needed to ensure that research funds are substantial and sustainable. Creative methods of funding should be considered, such as formula funding to establish new bases of support. In addition, it should be recognized that many important scientific discoveries are the result of work performed on the periphery of another funded project. Research sponsors should accept and encourage projects that promote exploratory studies in addition to more specific projects.

ASSEMBLING AN EXPLORATORY RESEARCH AND DEVELOPMENT PROGRAM

A preliminary step in designing an exploratory research program to anticipate environmental problems should be to determine the reasons that environmental problems have often gone unnoticed and how early identification and resolution can be achieved. The following steps were developed to help gain an understanding of environmental problems:

- Analyze case histories related to the management of significant environmental problems. The program should include efforts to synthesize the results of past efforts at environmental management. It should include studies of the whole process from risk identification to action and resolution. The 1986 National Research Council book, *Ecological Knowledge and Environmental Problem-Solving: Concepts and Case Studies,* is an excellent prototype. Such a study could focus on environmental problems resulting from chemical releases to the environment as exemplified by the chapters on "Ecological Effects of Nuclear Radiation" and "Environmental Effects of DDT."
- Scrutinize research programs at existing national and university laboratories, federal agencies, and independent research institutes to identify factors that have contributed to highly successful research programs. Similarly, exploratory research programs in industry should be examined for key factors such as institutional rewards for exploratory programs, methods of financing the research, incentives to the institution as well as to researchers.
- Evaluate the effectiveness, advantages, and disadvantages of programs carried out by individual investigators, research teams, and centers (such as those at NIEHS and its centers). Large, successful multidisciplinary research programs in areas outside of environmental sciences could also be scrutinized to learn what factors have contributed to the success of these programs.
- Establish unambiguous high priority and long-term commitment to environmental studies. The criteria for judging environmental managers would have to be changed to reward such long-term commitments.
- Foster and encourage expert scientific vision on a continuing basis by assembling experts to identify potential, emerging, or escalating environmental problems.
- Design international programs to exchange ideas, expertise, and experience in environmental studies. This international cooperation should be incorporated in the exploratory studies program. Historically, international activities and programs have provided valuable tools for learning about emerging problems. They have also provided unusual opportunities for studying pollutants that have affected more people and in much greater intensity than in the United States. Environmental problems in other countries can provide valuable lessons for anticipating and preventing similar experiences in this country, as well as providing a means of ensuring that other countries will not repeat the pitfalls experienced in the United States.

CONCLUSIONS

The general conclusion of the workshop was that future environmental problems will be solved only by a broad-based, long-term research program in addition to the short-term research and testing programs that provide the scientific foundations for current regulatory situations. Such a program will require well-designed environmental monitoring to establish baselines from which environmental changes can be measured and then followed. The program will need ongoing studies of the changing social situation, the public expectations for environmental management, and the outcomes of environmental interventions. New technologies need to be evaluated from two points of view: What new environmental problems may be generated, and what their contribution may be to alleviate environmental difficulties.

These goals will have to be incorporated into a long-term research program, which will provide career opportunities for first-rate professionals to develop a scientific discipline from natural and social sciences that provides the tools to manage the environment in the decades ahead.

Workshop Participants

Research Needs in Anticipation of Future Environmental Problems
Woods Hole, Massachusetts
June 14 and 15, 1989

Donald Hornig, Chair, Harvard University, Boston, Massachusetts
William Ascher, Duke Institute for Policy Sciences, Durham
Russell Christman, University of North Carolina, Chapel Hill
William Clark, Harvard University, Cambridge
James M. Davidson, University of Florida, Gainesville
Kenneth Dickson, University of North Texas, Denton
John R. Ehrenfeld, Massachusetts Institute of Technology, Cambridge
Peter Gresshoff, University of Tennessee, Knoxville
Peter Groenewegen, Rensselaer Polytechnic Institute, Troy, New York
Phil Gschwend, Massachusetts Institute of Technology, Cambridge
Robert Huggett, Virginia Institute of Marine Science, Gloucester Point
Roger Kasperson, Clark University, Worcester, Massachusetts
Joseph Ladou, University of California, San Francisco
Adrianne Massey, North Carolina Biotechnology Center, Research Triangle Park
Francois M. M. Morel, Massachusetts Institute of Technology, Cambridge
Stephen D. Parker, National Research Council, Washington, D.C.
A.J.M. Schoot-Uiterkamp, MT-TNO, The Netherlands
Keith Solomon, Canadian Center for Toxicology, Ontario
Barbara Walton, Oak Ridge National Laboratory, Oak Ridge, Tennessee
Irvin L. White, New York State Energy Research and Development Authority, Albany
Terry Yosie, American Petroleum Institute, Washington, D.C.
Dan Beardsley, EPA, Washington, D.C.
Carl Gerber, EPA, Washington, D.C.
Rick Linthurst, EPA, Washington, D.C.
James R. Fouts, NIEHS, Research Triangle Park

Appendix E

Commission on Physical Sciences, Mathematics, and Resources

Norman Hackerman (*Chairman*), Robert A. Welch Foundation, Houston
Robert C. Beardsley, Woods Hole Oceanographic Institution, Woods Hole
B. Clark Burchfiel, Massachusetts Institute of Technology, Cambridge
George F. Carrier, Harvard University, Cambridge, Massachusetts
Ralph J. Cicerone, National Center for Atmospheric Research, Boulder
Herbert D. Doan, The Dow Chemical Co. (retired), Midland, Michigan
Peter S. Eagleson, Massachusetts Institute of Technology, Cambridge
Dean E. Eastman, IBM T.J. Watson Research Center, Yorktown Heights, New York
Marye Anne Fox, University of Texas, Austin
Gerhart Friedlander, Brookhaven National Laboratory Associated Universities, Inc., Long Island
Lawrence W. Funkhouser, Chevron Corp. (retired), Menlo Park, California
Phillip A. Griffiths, Duke University, Durham
Neal F. Lane, Rice University, Houston
Christopher F. McKee, University of California at Berkeley
Richard S. Nicholson, American Association for the Advancement of Science, Washington, D.C.
Jack E. Oliver, Cornell University, Ithaca
Jeremiah P. Ostriker, Princeton University Observatory, Princeton
Philip A. Palmer, E.I. du Pont de Nemours & Co., Newark, Delaware
Frank L. Parker, Vanderbilt University, Nashville
Denis J. Prager, MacArthur Foundation, Chicago
David M. Raup, University of Chicago
Roy F. Schwitters, Superconducting Super Collider Laboratory, Dallas
Larry L. Smarr, University of Illinois at Urbana-Champaign
Karl K. Turekian, Yale University, New Haven

Staff

Myron F. Uman, Acting Executive Director